SMALLVILLE

SEASON 3

THE OFFICIAL COMPANION

Superman created by Jerry Siegel and Joe Shuster

SMALLVILLE: THE OFFICIAL COMPANION SEASON 3
1 84023 952 2

Published by
Titan Books
A division of
Titan Publishing Group Ltd
144 Southwark St
London
SE1 0UP

First edition September 2005
10 9 8 7 6 5 4 3 2 1

Visit our websites:
www.titanbooks.com
www.dccomics.com

Did you enjoy this book? We love to hear from our readers.
Please e-mail us at: **readerfeedback@titanemail.com**
or write to Reader Feedback at the above address.

To subscribe to our regular newsletter for up-to-the-minute news, great offers and competitions, email: **titan-news@titanemail.com**

A CIP catalogue record for this title is available from the British Library.

Printed and bound in Great Britain by MPG, Cornwall.

SMALLVILLE™

SEASON 3

THE OFFICIAL COMPANION

Paul Simpson

TITAN BOOKS

ACKNOWLEDGEMENTS

Thanks as ever are due to many people for their help in preparing this book. It wouldn't be possible without the cooperation of those members of the cast and crew who gave up their limited free time to be interviewed; Kendra Crowther in Vancouver, Lisa Rose and Susan Kesser in L.A.; and Chris Cerasi at DC Comics for coordination; Craig, Steve, April and the Deep Diver for all their help on and off the Web; Jo Cleary, Nicola Hickey, Andy Lane, and most especially Bev Simmons for morale-boosting; and Adam Newell, Rich Matthews, Jo Boylett, J. P. Rutter, Oz Browne and the team at Titan who continue to keep the faith.
— Paul Simpson

The publishers would like to thank the following people who contributed to the *Torch* and *Ledger* articles featured in this book: Jake Black, Angela Dai'Re, Christopher Freyer, Lisa Gregorian, Dan Manu, Karen Miller, Gena Murph, Will Rogers, Mimi Soo, Mark Warshaw and Kathryn Zoucha. The publishers would also like to thank Chris Cerasi at DC Comics, and the entire production team of *Smallville* for all their help with this project.

DEDICATION
This one's for Emma, who's always had faith.

KEEP YOUR FUTURE SAVINGS SAFE
SECURE
RETIREMENT
BANK OF
METROPOLIS

CONTENTS

INTO SEASON THREE

As the jet begins to spin, Lex scrambles up, clambers toward the cockpit. He swings open the door, stares into the cabin in stunned disbelief.

WHAT HE SEES: Both pilot seats are empty. The radio has been ripped out and the controls have been shattered.

At that moment, the plane bursts through the storm clouds. Lex peers through the windshield in horror as a RIBBON OF LIGHTNING illuminates

A MOUNTAINOUS EXPANSE OF OCEAN!

As the jet nose-dives toward the churning waves and a furious plume of lighting WHITES OUT THE SCREEN...

SMASH CUT TO:

A SCREAMING MOTORCYCLE.
It's speeding down...
EXT. COUNTRY ROAD - DAY

Clark sits on its back, wind whipping his hair, his eyes hidden behind mirrored sunglasses. As he powers over a rise, the CAMERA CRANES UP, HIGHER and HIGHER until the towers of METROPOLIS are revealed in the distance, gleaming like a high-tech OZ.

JOR-EL (V.O.)
You cannot escape your destiny, Kal-El.

As if in response, Clark guns the bike faster. As he careens toward this city of the future and an untold chapter in his saga, we...

FADE OUT.
END OF SEASON II

INTO SEASON THREE

"Season three will probably be our darkest season, until we get to later seasons and the Clark/Lex relationship is finally split once and for all."
— Al Gough

"The big themes going into season three were consequences, and Clark trying to run away from his destiny," executive producer Al Gough points out. "In terms of Lex's character, in our opinion, he takes a pretty big step toward the dark side. If you look at 'Exile', and then at 'Covenant', you see that all the themes set up in 'Exile' are played out throughout the season. The season starts off with Clark 'on drugs' in Metropolis, and coming back and dealing with the fallout, finally pushing Lana away once and for all. He becomes very isolated in season three... we're trying to put those pieces back together. There are a lot of darker tones than you saw in seasons one or two. It's probably the most emotionally consistent season from episode to episode."

Opposite above: The brand is burned into Clark's chest.

Opposite below: Jonathan Kent seeks help from Jor-El.

Below: Kal makes an unauthorized withdrawal.

The producers came into the season with plans for their lead characters, though the actors themselves weren't always aware of them. "We try not to tell the actors too much, unless it's helpful and will inform their performance in a good way," Gough states. "Otherwise we try to keep them in the dark as much as possible. We have a meeting with them at the beginning of the year and talk about bigger ideas and bigger themes for the season, where the character is going in a general sense, but in the specifics, they read it in the script. Sometimes we'll tell one actor something and say it's their secret, and they mustn't tell any of their castmates."

Season three promised the arrival of elements from Clark's future, as well as new superpowers, and a few answers — but as ever, would they lead to new questions? "To people who ask, 'How does *Smallville* end?' Clark doesn't end up with Lana, and he and Lex are enemies," Gough explains. "That is the thing with this part of Superman's life: it's a tragedy and it ends badly. The prequels are always the tragedy!"

As the cast and crew got back together after the summer hiatus, audiences were being prepared for the return of the show in the fall by a series of short commercials for the cell phone company Verizon, which showed some of Clark's exploits in Metropolis. But no one watching would have been prepared for the complete change in the lead character revealed at the start of 'Exile'...

THE EPISODES

"If I had the answer to half the 'whys?' in this town, I'd have a best-seller on my hands."

— Sheriff Nancy Adams

SEASON 3 REGULAR CAST:

Tom Welling (Clark Kent)

Kristin Kreuk (Lana Lang)

Michael Rosenbaum (Lex Luthor)

John Schneider (Jonathan Kent)

Annette O'Toole (Martha Kent)

Allison Mack (Chloe Sullivan)

Sam Jones III (Pete Ross)

John Glover (Lionel Luthor)

EXILE

WRITTEN BY: Alfred Gough & Miles Miller DIRECTED BY: Greg Beeman	GUEST STARS: Rutger Hauer (Morgan Edge), Ryan Robbins (Louis Leery Jr.), Terence Stamp (Voice of Jor-El), Jill Teed (Capt. Maggie Sawyer), Emmanuelle Vaugier (Dr. Helen Bryce), Brock Johnson (Edge's Thug)

DID YOU KNOW?

The choice of Metallica as a soundtrack for Kal's escapades in Metropolis was one of those "Hell, yeah!" moments for music supervisors Madonna Wade-Reed and Jennifer Pyken.

Three months after leaving Smallville, Clark, now calling himself 'Kal', is maintaining a lavish lifestyle, funded by robberies, in Metropolis. This brings him to the attention of crime-lord Morgan Edge, who wants him to break into LuthorCorp and obtain a small container. Only Chloe knows Clark's whereabouts, but has kept it secret.

The scar on Clark's chest occasionally glows painfully, and the only relief he can get is when he removes the red kryptonite ring. He attends Lex's memorial service and is spotted by Lana. Chloe admits where Clark has been living. Lana tries to reason with Clark, but he catches her calling his parents.

Lex is on a desert island with a mysterious stranger called Louis, who tries to convince him that Helen was responsible for the plane crash. Lex refuses to believe him and they fight, only to be interrupted by a fishing vessel that has seen Lex's fire signal. In reality, there is no one on the island except Lex. Lionel is furious when Helen turns up at Lex's memorial service and warns her he will avenge Lex's death.

To bring Clark home, Jonathan makes a deal with Jor-El. The Kryptonian gives him powers equal to Clark's. Jonathan interrupts a burglary by his son at LuthorCorp and the two Kents plummet out of a sixtieth story window.

MARTHA: The fact is, Clark doesn't want to be found. He'll come home to us when he's ready. We just have to have faith.

The third season of *Smallville* started with a bang, literally, as the red kryptonite-infected Clark caused mayhem in Metropolis. "It was a really good episode to start the season," executive producer Miles Millar recalls.

"The stuff on the motorbike was really fun to do," Kristin Kreuk laughs, and Tom Welling admits he enjoyed the chance to stretch his acting muscles as Kal, as well as the opportunity to get out of the studio to film the various scenes set in the city center. "Coming back and starting off in Metropolis was a lot of fun," the actor notes. "Production-wise we were out of the studio, downtown [on location in Vancouver], doing different things from what we normally do. Add to that the effect of the red K, and I had the chance to do something that I had never done on *Smallville* before. There's the whole scene where Kal goes to buy the car, and I could play that confidence, being relaxed and in control. That set the tone for at least the first couple of episodes, where Clark was not just going to get to do different things, but in the long run do things that he wasn't necessarily going to forget. He still learns from those experiences once he gets

Opposite: Clark is torn in two directions.

phone

Smallville Ledger

LEX LUTHOR HONORED WITH MEMORIAL

Three months ago, Smallville's world-famous, jet-setting playboy took himself off the market, breaking hearts around the globe. Our own crown prince, Lex Luthor, married Dr. Helen Bryce. The storybook romance of the dashing young millionaire and his physician bride took a tragic turn shortly after the couple exchanged vows. The pilot of the Luthors' plane parachuted out of the aircraft, leaving the newlyweds to fend for themselves. The bride survived. And now it's official — the groom perished.

Luthor's grieving widow, Helen, returned to Smallville recounting a honeymoon nightmare featuring a selfless act of heroism from her husband. She remembers sharing a champagne toast with her groom on the jet. The next thing she can recall is awakening in time to see the pilot grab their money and jump out of the plane. With only one parachute left, Luthor valiantly offered it to her, and he went down with the crippled craft.

Luthor's father, Lionel, was forced to call off any further search-and-rescue efforts and face the horrible truth — his son is no longer alive. Services for Lex Luthor took place yesterday in Metropolis.

Lex Luthor is survived by his wife, Dr. Helen Bryce, who currently resides in Smallville, and his father Lionel Luthor, who lives in Metropolis. Luthor's half-brother, Lucas, lives abroad. Luthor was preceded in death by his brother, Julian, and mother, Lillian.

Preferred memorials should be sent to the Smallville Food Bank.

By Kathy Romita

off the red kryptonite. It was an exciting thing to come back and do, as an actor, and it was an exciting way to come back into the season for the audience. I always like to play that other side of Clark from time to time — he does everything he wishes he could do, but knows better than to do. Red K is always fun!"

"Clark definitely comes back from his experiences a changed person," Miles Millar agrees. "That's what's great about the show: he starts as a boy and becomes a man. In that episode you saw a big leap in terms of personality and his character, with his darker side. That experience scarred him emotionally."

Writer Mark Verheiden adds, "The dark side definitely took over because the stuff he does is so wildly criminal, but the good side was always trying to bust out. Clark is an inherently good person — and that's the difference between him and Lex."

The casting of Rutger Hauer as Clark's opponent added a further touch of class to the episode. "We basically called his agent and said that we had this part, Morgan Edge, from the DC Comics universe, who would be in two episodes, with the potential to come back," executive producer Al Gough explains. "We spoke to Rutger directly and said that it was the season opener, which would be heavily

SMALLVILLE MUSIC

'Frantic' by Metallica
'Good Talk' by Shocore
'Hey Mama' by The Black Eyed Peas
'Calling All Angels' by Train
'All the Stars' by eastmountainsouth

Above: Lex's presumed death makes the headlines.

promoted, with a lot of exposure. He liked the idea that he could come back. On a practical level, the show is about Superman, and for an actor like that, it exposes him to a younger audience. *Blade Runner* was twenty-five years ago now."

LOUIS: Why are you in such a hurry to get back to a family that twisted?
LEX: Because my life is not here.

The scenes with Lex Luthor and his imaginary companion Louis on the island were, as Al Gough admits, "the biggest buy of the episode. What we wanted to set up was mental illness, the schism that he had, which Lionel later exploits. Even in the realm of *Smallville*, Michael playing the *Fight Club* gag, on the island talking to himself, was one of the most out-there ideas that we've done. In hindsight, if we had to do it again, I don't know if we'd do the same thing, but Michael pulled it off. It's a testament to his performance. Once the audience bought off on that, I knew that it was only going to get stronger in terms of where we were going with Lex this season."

John Glover remembers Lex's funeral scene and the confrontation with Helen very clearly. "I knew Lex wasn't dead, of course, when we shot that," the actor notes. "But I knew that I had to play that scene as if Lionel thought he was. I think it worked pretty well. Lionel knew Helen was the villainess, the reason that Lex was gone — the 'grieving widow' act didn't work with him!" ■

DID YOU KNOW?

Rutger Hauer had to translate the script into Dutch, and then memorize the lines in English in order to deliver his performance as Morgan Edge.

PHOENIX

WRITTEN BY: Brian Peterson & Kelly Souders
DIRECTED BY: James Marshall

GUEST STARS: Rutger Hauer (Morgan Edge), Brock Johnson (Edge's Thug), Terence Stamp (Voice of Jor-El), Emmanuelle Vaugier (Dr. Helen Bryce), Françoise Yip (Dr. Lia Teng), Link Baker (Pilot #2), Kurt Evans (Edge's Thug), Dan Payne (LexCorp Airport Security)

DID YOU KNOW?

As a birthday present for John Glover, Rutger Hauer arranged for a special low chair to be delivered to the set for him to sit on between takes, rather than the normal director's chair.

Jonathan persuades Clark to destroy the red kryptonite ring, curing Clark and removing his own powers in the process. They return to Smallville, where the Kents are selling the farm. Lex confronts his father, and then faces Helen, who says she panicked. He visits the Kent farm and is reunited with Clark.

Lionel tells his childhood friend Morgan Edge to get the vial of blood back. Edge traces Clark to Smallville, but Clark warns him off. Edge takes the Kents hostage, demanding the vial, but Clark has destroyed it. However, he gives Edge some fresh blood, and Edge tells Lionel that he can supply the source.

With kryptonite taped to his chest to keep him subdued, Clark is taken to the Metropolis docks. He manages to free himself, and destroys the truck he's been held in, making Lionel think Edge was planning to double-cross him. Edge dives into the water to get away.

Helen eventually admits her guilt, but escapes. Lionel congratulates Lex on his handling of the situation, and Lex says he wants to come and work for his father. Lex buys the Kent farm and hands the deed over to the Kents.

Clark tells Lana that he doesn't think a relationship between them could work.

LIONEL: I had search parties scouring half the world for you.
LEX: Looks like you picked the wrong half.

"The last scene of 'Exile' and the start of 'Phoenix' in many ways define the season," Al Gough comments. "The fight between Clark and Jonathan, and the deal with Jor-El really was the main through-line for the Kents during the year. It led to the heart attack, to Jonathan's strange behavior, and then when the deal is revealed by Kara in 'Covenant', it comes full circle."

The nature of the deal was kept secret from John Schneider. "It showed that come hell or high water, Jonathan was going to get Clark back," Schneider points out. "For a long time, we didn't know what sacrifice he had to make in order to get those powers."

"That was one of the cooler gags we did," Tom Welling's stunt double Christopher Sayour says. "300 feet from the top of the building to eight feet from the ground, head-first. We shot it in Surrey, off one of the government buildings there." There was some debate whether to do the stunt as a computer effect or with real stuntmen, but producer Bob Hargrove pushed in favor of "the real deal", as Sayour calls it. "It was one of the bigger gags going on in town."

The fight between the two superpowered Kents was fun to shoot as well. "On a

Opposite: Clark reluctantly pushes Lana away.

DID YOU KNOW?

The matte paintings establishing Metropolis are not based solely on Vancouver, but include buildings from other modern cities, including Singapore and parts of Hong Kong.

normal feature that would take a month, easily," Sayour says. "We pulled the whole thing together in two days. It was a stunt marathon. Tom and Schneider did a ton of the stunts themselves."

"We did more than we probably should have," Tom Welling laughs. "Obviously we didn't do the stunts with the big impact, but we came up with quite a lot of ideas while we were doing it. They wanted to show our faces doing it, so whenever you can see it's us, it really is us. I think it adds to the intensity of the scenes, and gives more to the viewing audience. I do as much as I can, but of course if anything were to happen to me, there's a lot more at stake than just my health."

At the end of the fight, Clark's Kryptonian tattoo disappears from his chest. "We've done healing effects before, and we were able to get a really great-looking effect for the symbol disappearing," visual effects supervisor John Wash recalls. "It brings to mind time-lapse stuff like flowers growing, with that complexity of development, but running in reverse."

"Joey Bratazani and Brian Harding worked on those shots," adds visual effects producer Mat Beck. "It's a deceptively simple shot, because you have to track it believably on a moving, performing human chest. It can't just be this spot of light: it has to look like skin being backlit from burning underneath. It combines a certain amount of impact with performance, and it looks like it hurts!"

Below: Morgan Edge tracks down Clark with a copy of the Kents' homemade poster.

Smallville Ledger

LEX LUTHOR LIVES!
But Now Wife Is Missing

In an incredible twist of fate this week, Lex Luthor returned from a watery grave days after his own funeral, only to narrowly cheat death once again, losing his wife in the process. The stunning chain of events was a whirlwind of triumph and tragedy befitting the epic history of the area's most prominent and powerful family.

Although official details have not been released, pending a full investigation, a source within the Federal Aviation Authority said it appeared the plane's pilot suffered a sudden loss of consciousness that sent the plane plummeting from the sky. The cause of the pilot's lapse and his current condition are unknown at this time.

Displaying incredible instincts yet again, Lex was able to take control of the aircraft's throttle and stabilize it. However, because the accident seemed to have caused the plane's door to blow open, the sudden loss in cabin pressure may have sucked Dr. Bryce out of the plane, although a conflicting report from one source says that one of the plane's parachutes was missing.

After miraculously piloting the plane to a safe landing on a nearby island, Lex immediately called for a search for his wife to begin in the surrounding waters, despite the odds being against her surviving such a fall. That search is still underway, with Lex expected to devote all of the resources available to him to finding his wife.

By Frank Moore

Mark Snow altered the musical tone of the episode to fit the more somber tones required for Metropolis. "With Rutger Hauer as the villain, it felt like it needed a 'thick' darkness," he notes, "maybe not so much in tempo, but like a lot of John Barry's music — slow and ponderous, with deep sounds and a sense of importance."

CHLOE: You can't keep running away, Clark. We all do things we regret. Sometimes you have to stop and face your demons.

John Glover enjoyed his confrontation scenes with both Rutger Hauer and Michael Rosenbaum in this episode. "Rutger and I had a great time together," he says. "We were rivals but childhood friends. We ran out of time to shoot the look I wanted to give at the docks when I thought he had been killed, as if to say, 'Where's my friend gone?'"

The moment when Lex and Lionel come face-to-face for the first time in three months was very powerful. "We had fun that day in the office," Glover says. "It was nice, wonderful, honest emotion between a father and son who were distrusting of each other and yet needed each other. It seemed to have a nice complex emotion to it." ■

SMALLVILLE MUSIC

'White Flag' by Dido
'Maybe Tomorrow' by Stereophonics

EXTINCTION

WRITTEN BY: Todd Slavkin & Darren Swimmer
DIRECTED BY: Michael Kettleman

GUEST STARS: Harris Allan (Jake Pollan), Jesse Metcalfe (Van McNulty), Camille Mitchell (Sheriff Nancy Adams)

DID YOU KNOW?

Van McNulty's father was the Marine killed by Tina Greer when she was disguised as Whitney in season two's 'Visage'. It's never explained onscreen how Van knew that Tina was responsible.

Jake, a kryptonite-infected boy who can breathe underwater, attacks Lana in the Smallville High swimming pool. Her life is saved when Jake is killed by an unseen sniper, who leaves a meteor rock with the word 'freak' carved into it beside the body. Clark and Chloe discover that Jake had gills, and that he was a friend of Van McNulty, another student. McNulty's father was killed by Tina Greer (in the episode 'Visage'), and McNulty is taking revenge, using information he has hacked from Chloe's computer. Chloe has maintained a file on Lex, and Clark prevents him from becoming McNulty's next target. Seeing the effects that meteor rocks have on Clark, McNulty makes a kryptonite bullet to use on him. After mistakenly thinking he has killed him, McNulty faces Clark in a final confrontation at Smallville High, where Clark uses a bulletproof vest to save himself from the deadly kryptonite projectiles.

Lex wonders if the meteors affected him, since he has an elevated white blood cell count and has never been ill since the meteor shower. Chloe tries to get out of her deal with Lionel Luthor to report on Clark, but Lionel threatens her father's job. He also warns her not to investigate the Luthors again, or she will regret it.

CHLOE: Since when is curiosity a crime?
CLARK: Since someone took your database and turned it into a personal hit list.

Many of the ideas that materialized during the third season of *Smallville* had been discussed during the very early days of production on the series. Some had been put on hold for practical reasons — there being a limited number of episodes within each season, certain ideas were more time-sensitive than others, while others were waiting for just the right moment. The kryptonite bullet fired by Van McNulty was one of these.

"We'd had the idea during season one, but could never figure how to do it," Miles Millar recalls. "'Extinction' is a really solid episode of the show — it's a really good fit, a stand-alone, cool episode."

"It was one of those 'What if?' scenarios that we regularly discuss," Al Gough adds. "It was a fantastic effect, and it was great watching Clark's face as they shot him, and watching Jonathan take the bullet out."

"We developed a neat look for the effect," John Wash says. "We have a lot of effects where we slow down time, and we show bullets flying toward Clark, and he either dodges them or catches them, or bats them away, so we were wondering what we could do to make this a little more interesting. We had the idea that kryptonite might have a different reaction with the atmosphere as a bullet traveling at high speed. So we developed a little green aura and a little plasma envelope around the bullet as it's

Opposite: Van McNulty prepares to shoot Clark.

MUSIC

'Way Away' by Yellowcard
'Give It to Me' by Vishiss
'I Give You Take' by Maria
'Goodbye Again' by Vertical Horizon

SMALLVILLE TORCH

ANATOMY OF AN ASSASSIN

Smallville High School student Van McNulty allegedly went on a recent killing spree, murdering fellow student and swimming phenomenon Jake Pollen here on campus. McNulty is now also linked to the death of a local auto mechanic and the attempted murder of businessman Lex Luthor.

When the elder McNulty, a veteran Marine working as a recruiter in Smallville, died at the hands of Tina Greer last year, young McNulty's quarry shifted. The orphaned son's thoughts burned with revenge, and regrettably, he may have found fuel in these very pages. When the shape-shifting Greer fell to her death on a wooden spike last year, the *Torch* suggested a correlation between Greer's abilities and the meteor shower.

Torch investigations uncovered a hit list at the McNulty hunting cabin that included the names of both Wallace and Pollen along with others, including Lex Luthor. Autopsy data suggest Pollen possessed abnormal aquatic abilities, which might explain his underwater attack on Lana Lang. Out-of-the-ordinary behavior on Wallace's part is rumored but unproven. His family declined an interview. Whether McNulty's intended victims were connected by the meteor shower or not, McNulty believed they were. Before his arrest, McNulty took Lana Lang hostage. Lang said McNulty left no doubt as to his motives. "[McNulty] said, 'These freaks aren't people, Lana, no matter how much they may seem like it. Look at Jake. He was my best friend for three years, and I had no idea.'"

Van McNulty is now in custody. A trial is pending.

By Clark Kent

coming toward Clark. People would instantly know that it's a lot more exciting than the normal bullet shots that we do."

Filming Van McNulty's first freak attack required Kristin Kreuk to spend longer than she would have preferred in the water. "'Extinction' was about thirteen hours in the pool," she remembers. "It was pretty rough. I had fun for a while because I love swimming — it's one of my favorite things to do — but your eyes do get bloodshot after that length of time."

Al Gough also liked the episode because "it wasn't about a meteor freak. It was about somebody out there taking out meteor freaks because their father had been killed by one. I thought it gave a unique perspective — there is an awareness beyond our merry band that these people exist, and people have certain attitudes about them. I think it forced Clark, Lana, and Chloe to deal with their issues about them — it's an episode about tolerance. The guy finds out that Clark has got superstrength, so he debunks the guy using a lead vest to deflect the kryptonite bullets."

The confrontation between Lionel and Chloe where she resigns was originally scripted to appear in 'Phoenix', rather than 'Extinction'. "We pulled the scene out of there, and put it in 'Extinction', and it worked better in that episode than it would have

done in 'Phoenix'," executive producer Ken Horton comments. "She and Clark have patched up, and she goes from there to see Lionel. If I bring anything to this party, I have good vision for that sort of thing. Most people wouldn't go, 'Hey, let's move a scene from one show to another.'"

Above: Clark and Jonathan debate their next move.

LANA: Clark, if you don't open up to the people that love you, you'll always be alone. I can't believe you want to spend your life like that.

According to producer Mark Verheiden, there was never much intention to keep Chloe working alongside Lionel for very long. "We wanted to change Chloe's character a little bit, in a darker way, but we decided finally that it didn't feel right for Chloe to be sabotaging her friend Clark for Lionel," he notes. "In 'Exodus', at the end of season two, she was all in black, and we didn't want her to become the Dragon Lady. We didn't want to go there with her. She is one of Clark's dearest friends, and to have her constantly undermining him, she'd almost have become mini-Lex, and we don't need someone else doing that." ■

DID YOU KNOW?

Jesse Metcalfe (Van McNulty) is also well known for his role as John Rowland on *Desperate Housewives*.

SLUMBER

WRITTEN BY: Drew Greenberg DIRECTED BY: Terrence O'Hara	GUEST STARS: John De Santis (The Traveler), Katharine Isabelle (Sara Conroy), Christopher Shyer (Nicholas Conroy)

DID YOU KNOW?

The big red costume for the Traveler was one of the favorite creations of the costume department for the early part of the third season.

Clark wakes from a vivid dream to discover that he has been asleep for thirty-six hours. In his dream, a girl named Sara Conroy asks for his help against a huge red-cloaked figure called 'The Traveler'. Sara is in fact the girl whose uncle has just moved into Lana's old house. Clark and Lana learn that she has been in a coma ever since a car accident six years ago.

At Sara's house, Clark spots a cabinet of drugs that her uncle might be using to keep Sara comatose, and Lana learns that Sara's uncle controls her fortune while she remains unconscious. Clark keeps falling asleep and trying to help Sara. As Clark battles the Traveler in the dreamscape, while persuading Sara to fight against him, Lana goes to the Conroy's house, where she is captured by Sara's uncle. He is about to kill her when Clark awakens and comes to her rescue. Without the drugs, Sara wakes up and tells Clark that she knows Lana is the girl of his dreams.

Lionel wants Lex to undergo a series of psychological tests before he gives him full access to LuthorCorp. Despite initial resistance, Lex agrees.

LEX: I'm going to dedicate myself to ensuring the whole world knows who Clark Kent really is. Life as you know it is over.

As Tom Welling points out, the first act of 'Slumber' is designed to throw the audience. "You think, 'Wait a minute — did I miss a whole season?'"

"From a storytelling point of view, we wanted to change it up," producer Mark Verheiden recalls. "We really wanted to go for this idea that the entire first act is in fact a dream, and that all the strange stuff that's happening adds up. We didn't want to make it too strange, because we really wanted the audience to go along with it. They needed to feel that they were in Clark's world, but things are odd. Once we got to the point where Lex whacked him over the head, people hopefully realized that this was some sort of alternate reality."

"We tried something different," executive producer Al Gough agrees. "It's one of those episodes where you swing for the fences, and it has some very memorable scenes. I don't know that the format of television, with a commercial break between the teaser and the first act, necessarily lends itself to suspending that disbelief, but it had the great Clark/Lana skinny-dipping scene, and Lex breaking the sword on Clark. It was a little out there even for us, but if you don't try those episodes, you're never going to know. I think it was a definite tilt into fantasyland — it was so theatrical. When it aired, the response was that people either really liked it or really hated it. People had an opinion,

Opposite: Clark tries to deflect Lex's attack.

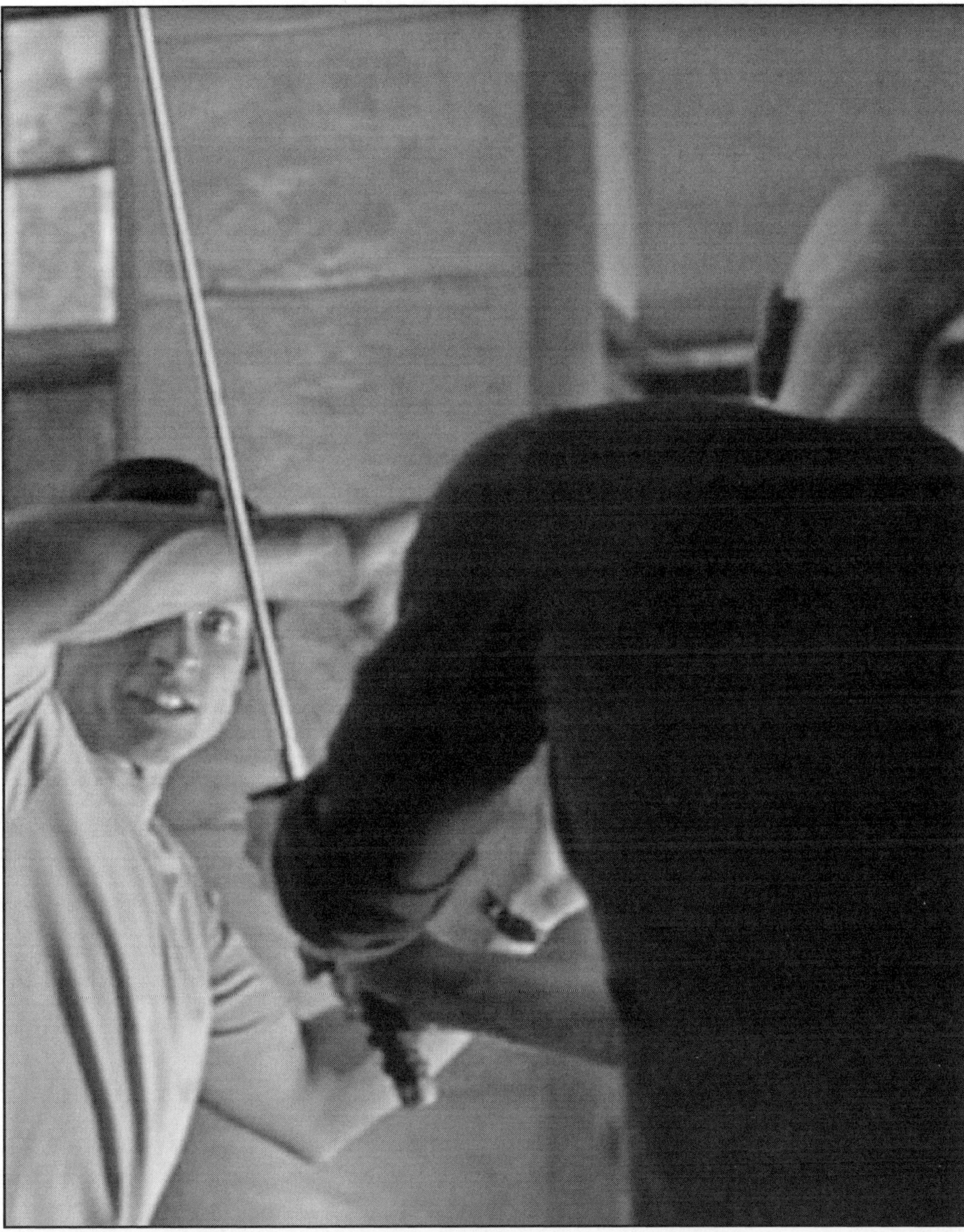

SMALLVILLE MUSIC

Music: (all by R.E.M.)
'Imitation of Life'
'Everybody Hurts'
'Losing my Religion'
'Bad Day'
'At My Most Beautiful'

rather than just saying it was okay."

Tom Welling is definitely one of those who enjoyed it. "'Slumber' may not really touch much on the big [story] arcs for Clark, but any time that you can touch on those arcs between Lex and Clark, I think they make the most interesting episodes," he points out. "You know that this is just setting something up which is going to be huge at some point years down the line."

One of Tom's favorite visual effects in the series is the attack by Lex on Clark, during what the audience later realize is Clark's dream sequence. "The sword is right there, and I'm blocking him," he recalls. "You can see Lex, and the sword just shatters perfectly. It's a great visual effect."

Michael Rosenbaum enjoyed the opportunity "to go overboard. It was one of those classic moments — 'You should have trusted me. I was your friend, Clark, why did you let me down?' That was really a true moment, even though it was a dream. It really was what would have happened if Lex had found out. 'How could you do this? Why didn't

Below: In his dreams! Clark and Lana get together at Crater Lake.

SMALLVILLE TORCH

REBUTTAL

Ever since we ran Clark Kent's article, "Anatomy of an Assassin", readers have flooded the *Torch* office with letters decrying our "journalistic integrity" and not so subtly suggesting our — but mostly my — responsibility in the Van McNulty murders.

Here are some of the more colorful snips from the letters we've received.

"I'm captain of the swim team. My coach tells me I have 'abnormal aquatic abilities.' Can I get on your hit list, too?" senior Scarlett Friedman wrote.

An anonymous reader wrote, "If the next student murderer has any brains, he or she will take their victims not from your articles, but from the detention rolls or the student court blotter."

Our coverage on meteor-effect phenomena aims not to vilify or venerate those affected by it but to represent the truth. That is why this paper will continue its investigations to these ends.

Van McNulty's heinous acts underscore why this town must find the answers behind its defining moment. As long as the meteor shower's effects remain a mystery, the opportunity for violence precipitated by ignorance persists, and the unknown dangers to our town's population lie covered by cornstalks, submerged in farm ponds and, most dangerously, buried in apathetic consciences.

By Chloe Sullivan

you level with me? I was your best friend. All I did was give to you — why did you lie to me?' Lex felt devastated by that."

Allison Mack laughs at some of the dialogue she had. "My character always has these [amazing] lines," she says. "Chloe says, 'Telepathic activity is increased during REM sleep.' Who talks like that? But I have to come out and say it, and act as if it's something that I've known for years. It's like playing a doctor, coming out with this technical jargon."

LEX: Clark, from all appearances the guy hasn't done anything wrong.
CLARK: We both know appearances can be deceiving, Lex.

Bearing in mind the nature of the episode, REM couldn't have been more appropriate a choice for the music. "Ken Horton had wondered if there was a band with whom we could do an entire episode," music supervisor Madonna Wade-Reed recalls. "Jennifer Pyken and I were having breakfast with Warners [Music], and asked if they had anything coming out. They said they'd got REM, and all the eyes lit up around the table. Everybody got copies of the album and weighed in on what songs they wanted. Certain of the old REM favorites are so timeless that even a new generation of people being introduced to them could appreciate them, and they don't show their age."

DID YOU KNOW?

John De Santis (The Traveler) played Lurch the butler in *The New Addams Family*. He also appeared in *Dead Like Me*, as a character called Tiny!

PERRY

WRITTEN BY: Mark Verheiden
DIRECTED BY: Jeannot Szwarc

GUEST STARS: Michael McKean (Perry White), Lorena Gale (Dr. Claire Foster), Camille Mitchell (Sheriff Nancy Adams)

DID YOU KNOW?

Annette O'Toole auditioned for director Jeannot Swarzc in 1972 for a TV movie called *Lisa Bright and Dark*.

While Clark's ability to control his powers is severely affected by a giant solar flare, he meets Perry White, a formerly top-class reporter who has fallen upon hard times. He's investigating the strange phenomena around Smallville, and persuades Clark to be his guide. White antagonizes Lana, so Lex runs him out of town. However, his interest in Clark is piqued when a tractor falls out of the sky in front of him, thrown there by Clark in an accidental fit of superstrength. White tries to enlist Lex's aid for information on Clark in return for material on Lionel, but Lex warns him off. White then threatens to commit suicide unless Clark saves him, but Clark's powers have temporarily disappeared. White jumps into a gorge with only a rope around his ankle and Clark dives after him, leaving Pete and Lana to come to their rescue. White realizes he has made a fool of himself, and vows to give up drinking and try to get his career back on track. He tells Clark to look him up in Metropolis someday.

Lex's sessions with psychiatrist Dr. Claire Foster aren't going as well as he hoped, as she insists on delving deeper than he would like, but eventually he agrees to cooperate, on condition of complete confidentiality.

PERRY: I made exactly two mistakes in my life, kid. The first was getting into journalism.
CLARK: What was the second?
PERRY: Thinking it mattered.

"'Perry' is one of our lighter episodes," writer Mark Verheiden notes. "After the heaviness of the first two episodes, followed by 'Extinction' and 'Slumber', we needed a little light relief. It's in the vein of season two's 'Heat', but is even funnier."

For all its comedy, 'Perry' served a number of purposes within the *Smallville* mythology. "We finally answer that age-old question, 'Why did Perry White hire Clark Kent?'" executive producer Al Gough notes. "And Clark discovered that all his powers come from the sun. It's a terrific episode — we had Michael McKean, and casting is ninety-nine percent of the episode. He did a terrific job, and locked into that character. We saw the man at his nadir, and by the end of the episode, he was back on the path to becoming the great journalist and editor that we know he'll become."

Opposite: Don't call me Chief — yet! The soon-to-be-feared Perry White.

"It's a different take on Perry," Mark Verheiden adds. "It's suggesting that he had a bit of a journey before he became the editor of *The Daily Planet*. He was a reporter who ran an exposé on Lionel Luthor, and ended up with some information that Lionel

SMALLVILLE MUSIC

'Walking in Memphis' by Lonestar
'Blue' by LeAnn Rimes
'I Love This Bar' by Toby Keith

decided he didn't want him to have, so Lionel essentially ruined him. This Perry didn't have the intestinal fortitude or strength to fight that yet. What happens in the course of the episode is that by meeting Clark, he ends up finding that strength. DC Comics has been really great about giving us a certain amount of latitude with these characters. We always keep in mind that we're not out to mangle any continuity for Perry White. He can still be the feared and revered editor that he will become. I would be surprised if a guy as irascible and cranky as Perry White didn't have a few bumps in the road. That's the type of guy he is. He's gone through a little bit more pain than you might guess from the Perry of the comics or the movies. We gave more of a sense of depth to that trip to becoming editor."

Michael McKean, who is married to Annette O'Toole, was a popular figure on the set. "He just *was* Perry," Tom Welling notes. "More than anyone else I've seen, you can't see the acting. You don't see any separation — you just see this character. Whatever he did in preparation in terms of his skill and talent, he just is Perry. In an odd way, there was a family feeling already, because we are somewhat of a family up here in Vancouver."

The moment at the end of the episode when Clark throws a bale of hay into the loft and catches it himself looks seamless on screen, but involved a number of effects. "Tom throws the bale of hay up, which is on a wire which we remove," Mat Beck explains. "Then he moves up into the loft at normal speed, which we speed up, and then when he's standing there, he catches a bale of hay that's thrown by a crewmember. We take

Below: A sunspot weakened Clark can't lift the tractor... initially.

those pieces, put them together, and make it look like it's all the same bale of hay. Taking the time to make the bits marry up correctly is very important."

'Perry' saw the introduction of Lex's psychiatrist, Dr. Clare Foster, played by Lorena Gale, who was very impressed with the episode's director, Jeannot Swarzc. "He was very precise about every movement," Lorena notes, "and in taking that time, he didn't have to use six takes to get what he wanted." Annette O'Toole agrees. "I adore this man," she says. "He was just phenomenal. The crew adored him, too." "It was a blast getting the dailies, and seeing how well it was coming together," Mark Verheiden adds.

DID YOU KNOW?

By a weird coincidence, there really was increased sunspot activity during the week that this episode was first broadcast in the United States.

CHLOE: Wise decision coming inside. The weather service expects a light shower of threshing machines followed by a drizzle of combines.

The episode was unusual in that a high proportion of the first draft of the script was used. "We barely changed it," executive producer Greg Beeman recalls. "Verheiden nailed it, and got the mythology and the humor."

"It all worked really well," Al Gough concludes. "That last image of Clark at the crossroads with Perry's bus heading off was terrific." ■

Daily Planet

METROPOLIS' GREATEST NEWSPAPER

DAILY PLANET SCRIBE NABS ANOTHER KERTH NOD

The Kerth Award board has released the names of the finalists for this year's awards, the most prestigious in all of print journalism. Included among the nominees announced yesterday was *The Daily Planet's* own Perry White, who was cited in both the investigative reporting and public service categories.

White was recognized for his part in uncovering the corruption scandal that rocked Metropolis City Hall earlier this year, resulting in the resignation of veteran City Councilman Brewster Boland.

In a series of front-page exposés, White revealed that Boland had apparently accepted cash bribes in exchange for favors to the developers of an industrial plant that endangered the health of residents in one of Metropolis' poorer neighborhoods.

Although Boland initially denied the allegations, White found and interviewed several eyewitnesses to the illegal payments. He also unearthed a paper trail linking the politician to an offshore bank account in the Cayman Islands.

Since resigning from the council in disgrace, Boland has been the subject of a federal grand jury probe as well as several civil lawsuits, all of which are still pending in the Metropolis courts. According to recent reports by White, Boland's legal fees are allegedly being paid by longtime acquaintance and one-time campaign contributor, LuthorCorp CEO Lionel Luthor. The company has declined to comment on the matter.

By Martin Exley

RELIC

WRITTEN BY: Brian Peterson & Kelly Souders
DIRECTED BY: Marita Grabiak

GUEST STARS: Peter Benson (Lachlan Luthor), William B. Davis (Mayor Tate), Tom Heaton (Dex McCallum), Tim Henry (Private Investigator Mason), John Mann (Young Dexter McCallum), Camille Mitchell (Sheriff Nancy Adams), Michael Karl Richards (Sheriff Billy Tate), Kendall Cross (Clark's Grandmother), Dean Wray (Hiram Kent)

DID YOU KNOW?

John Glover asked if he could play the part of Lachlan Luthor, but it was decided that Lachlan wouldn't be one of the doppelgangers.

In 1961, Dexter McCallum was convicted of murdering his wife — Lana's great aunt Louise — but now, knowing that he is dying, he contacts Lana to plead his innocence. He believes the real killer was a drifter who vanished shortly after the murder — whom contemporary newspaper reports show as Clark's double. Clark finds the medallion the drifter was wearing, which belonged to his father, Jor-El. Clark is then bombarded with images from 1961.

Jor-El rescues Louise from a mugger (who is arrested by Sheriff Tate), and they fall in love. 'Joe' reveals that he is an alien sent to Earth to learn humility. Their farewell is interrupted by the mugger, Lex's grandfather Lachlan Luthor, who shoots at Jor-El. The bullets can't harm the Kryptonian, but one ricochets and kills Louise. Dexter hears the shots and sees Jor-El run off, but is arrested for the murder by Tate. Clark's grandfather, Hiram Kent, helps Jor-El to escape.

Back in the present day, Clark scares Mayor Tate into admitting that he released Luthor to kill Joe. He is arrested and Dexter is freed. Clark thinks that Jor-El deliberately chose the Kents as his adoptive parents a result of Hiram's help.

Lionel admits to Lex that his parents died in a tenement fire, which Lex discovers might have been arson.

JOR-EL: Where I'm from we have colors that you've never seen. Our moons are so close they fill up half the sky. We have sunsets that last for hours.

"'Relic' was a tough episode to shoot, but it's an episode that people really like," Al Gough admits. "It was really saved in post-production by Ken Horton, and by Mat Beck and his crew. It was a terrific script."

In the early part of the season, the production team was consciously staying away from too many references to Clark's Kryptonian heritage, with the exception of this episode. "We decided that the first part of the year was more about him bringing himself back to being a high school student," producer Mark Verheiden recalls.

Like the kryptonite bullets eventually featured in 'Extinction', the idea of doppelgangers for the characters was another of the concepts thought of much earlier in *Smallville*'s history. "We talked about that early in season one," Verheiden says, "but we all felt that we needed to have more of an investment in the characters before we did that. The audience needed to know them better." The doppelganger idea allowed the

Opposite: Jonathan and Clark try to puzzle out the relic in the cave.

DID YOU KNOW?

Reshoots for this episode took place during the filming of 'Asylum'.

audience to connect closely with the story being told, filling in another piece of the mythology. "We see in this episode that Jor-El has been here before, and that Clark's landing in Smallville wasn't happenstance," Al Gough says. "There was a destiny and a plan, and Clark is somehow inextricably linked to the Langs, and specifically to Lana. 'Relic' played into all the core themes of the show."

Going back in time to tell this romantic story played well with the audience. "To this day, when people come to me, they talk about the 'back in time' show," Al Gough points out. Both Tom Welling and Kristin Kreuk enjoyed the chance to play a different character. "It was great to play Lana's great aunt, who was such a bitch!" Kristin says bluntly. "It was so much fun, and I think we pulled it off."

"I tried to portray a little more maturity in that character," Tom adds. "He didn't have so much compassionate understanding of the human race as Clark has, and Kristin and I were interested in establishing a relationship between the characters where you could hopefully see that they were destined to be together forever."

Tom wasn't that surprised that he was called upon to fly. "It would be insane to think that you could play Clark Kent and not get into it at some point," he says wryly. "But you have to look at each circumstance and how it affects the greater story arc, while still maintaining the integrity of what we first signed on to do — Clark Kent in high school.

Smallville Ledger

LOCAL BEAUTY MURDERED

A Community Shocked by Murder of Farmer's Beloved Wife

Louise (Potter) McCallum, 21, was shot to death last night in the McCallum family barn. Sheriff's Deputy Billy Tate arrested Louise's husband of two years, Dexter, for the murder. Tate heard gunshots as he was driving past the McCallum farm on his way back to the station. He immediately stopped to investigate and saw Dexter cradling his dead wife and holding a gun by his side.

Tate said, "It was a real foggy night last night, and I was returning to the station after patrolling the back roads for a while. I heard some gunfire while I was about fifty yards from the McCallum driveway. I pulled in and headed for the barn 'cause that's where the popping came from. And I found Louise dead. I immediately placed Dex under arrest."

Neighbors and friends of McCallum are shocked. Helen Romita said, "He loved Louise more than life itself. I just know they've got the wrong man. Dex keeps insisting he saw a stranger, a drifter, on his property. I bet the drifter did it."

Mrs. McCallum was born in Smallville and educated here. While attending Smallville High School, she was very active in theatre and starred in several productions. Her former drama teacher, Mr. Thompson, said, "Louise had 'it'. Star quality. It seemed to be her destiny. I always thought she could be in movies. This girl was ahead of her time." But Louise stayed in Kansas and married Dexter McCallum in 1959.

By Mitch Stone

Above: The sultry seductress Louise McCallum.

That character I will never allow to be Superman on the show, but it's okay to make it another character."

The episode also served to provide another side of Jor-El. Up to then, Jor-El hadn't seemed like a very pleasant person. "It's the classic case of the adopted child looking for his birth parents, finding them, and realizing that maybe they're not nice people," Al Gough explains. "But nothing that Jor-El has done up to now has been bad. He said to Clark, 'Come to me,' and Clark panicked. He blew up the ship, ran away, then came back. You've got to look at what Jor-El's done at this point, and realize that it's malevolent, but not outwardly evil."

CLARK: I think Jor-El was sent here as some kind of rite of passage. Grandpa Kent was down here, too. He told Joe that if there was anything he needed...
JONATHAN: What is it, Clark?
CLARK: I don't think you and Mom found me by accident.

To create the town of Smallville forty-two years earlier took a lot of work from the show's different departments, and they all rose to the challenge. "My entire department enjoyed working on 'Relic', in which we got to dress Lana as Louise," wardrobe mistress Caroline Cranstoun says. "It's always fun when we get to do something different."

Madonna Wade-Reed went to extra efforts to ensure that the music soundtrack was accurate. "I was really meticulous, and researched what songs were on the charts. Our music fans are really diligent, and I'm not out to pull a fast one on them. I gave myself about a four to six month window, and tracked down songs from that period that I thought would work." ■

SMALLVILLE MUSIC

'In Your Eyes' by Aaron D
'I Only Have Eyes For You' by The Flamingos
'Earth Angel (Will You Be Mine)' by The Penguins
'Matchbox' by Toby Keith

MAGNETIC

WRITTEN BY: Holly Harold
DIRECTED BY: David Jackson

GUEST STARS: Kevin Zegers (Seth Nelson), Tim Henry (Private Investigator Mason), James Hutson (MRI Doctor), Josh Kelley (Himself), Camille Mitchell (Sheriff Nancy Adams), Steven Cree Molison (Trucker), Rimple Sumer (Wendy Schwiebe), Douglas Chapman (Deputy)

DID YOU KNOW?

Guest star Kevin Zegers won the Young Artist Award for his portrayal of Josh Framm in the first of the three *Air Bud* movies.

Lana's lab partner Seth Nelson is affected by a snow-globe filled with meteor rock crystals. After an MRI scan, he realizes he has magnetic powers, which can move objects and alter a person's emotional state, and uses his newfound abilities to persuade Lana to date him. Clark is suspicious, but Lana tells him not to interfere. Seth makes Lana empty the Talon cash register and run away with him, but when they are stopped, Seth leaves Lana to face the sheriff on her own. Shortly after, he breaks her out of jail, and they hide in the fairground. Clark rescues Lana, rendering Seth unconscious, and although she apologizes for her actions, she also reminds Clark that she can date anybody she wants to.

Lex learns that Chloe is also looking into Lionel's background, and tries to warn her of the dangers. A detective working for Lex steals Chloe's computer, but is killed. Lex tries to persuade Chloe to work with him, and she admits that Lionel asked her to spy on Clark. She wanted to get an edge over Lionel, and Lex promises to protect her.

SETH: I love this ride
LANA: Yeah, me too. It reminds me that there's a world out there beyond Smallville. I'm going to leave one day.

Along with the episodes that deal with the various arcs that have come to characterize *Smallville* — whether it's Clark learning about his origins, or Lex's battles with Lionel — there are a number in each season that are stand-alone stories in which the villain of the week is introduced, his or her threat made clear, and then dealt with by Clark, without reference to anything other than the general effect that the meteor rocks have on the population of Smallville. During the first season, these 'Freak of the Week' stories met with some criticism from the audience, and from the second season onward, they have been less frequent. 'Magnetic' is a rare example of a third season story of that type.

"It may feel like a throwback," Al Gough agrees, "but it was one of our highest-rated episodes of the season, because it deals with a very specific teen issue. What happens if you become attracted to the wrong person, and they start to change you? It was one that the fans didn't like, but like 'Slumber' and 'Obsession', it was high-rated."

The executive producer notes that there's a fine balance that has to be maintained between the episodes that please the fans and those that cater to the network's primary audience. "You've got a teen audience on the WB," he points out. "Quite frankly, we got too mythology-heavy during the third year, and the audience went away." Episodes

Opposite: Seth tries to hold the police at bay.

SMALLVILLE SHERIFFS
STATION
DIRECTORY
SHERIFFS OFFICE
CELL BLOCK
TRAFFIC INFO
MAJOR CRIME
FINGERPRINTING
ASHROOMS
HAVE A NICE DAY

DID YOU KNOW?

Director David Jackson is an old hand at genre shows, with credits on *The Invisible Man* as well as *Lois & Clark: The New Adventures of Superman* to his name.

like 'Magnetic' pulled them in because the concept was clear from the trailers. "If it's an episode that people can plug into, or get some emotional grip on, they'll watch it. When you get into the secret lab episodes, it's just not relatable."

Kristin Kreuk found 'Magnetic' a major shift in pace from the high drama of 'Relic' the previous week. "I went to that from shooting 'Relic', which I was very excited about," she admits. "I didn't think that the way Lana behaved was part of who she was any more. Kevin Zegers was a nice guy, and he made doing that episode a bit more fun."

Allison Mack was pleased that it gave a rare occasion during the third year to show Chloe's lighter side. "She's the modern day Nancy Drew normally," she notes, "and during this year she's caught up in the intrigue and the mystery. But in 'Magnetic', you got a chance to see Chloe in a social setting with another woman. There's a tendency for her to show her immaturity, and the audience can see her as a teenager, because she's not figuring something out, or in love with Clark. She's being a young girl and having a good time. I think that's a really lovely part of her that you don't see that often, because she's silly, funny, and young. I always love those scenes with Kristin where I can be silly."

Although she did have some less dramatic scenes, the events of 'Magnetic' did start to demonstrate to Chloe exactly what she was getting into when she decided to take on Lionel Luthor. "There were some nice resonant moments where she learned about Lionel," Mark Verheiden comments. "There had been the moment in 'Perry' where he

Below: Lana and Chloe discover the kryptonite-filled snow globe.

tells her he was destroyed by Lionel, and Lionel will do anything to destroy someone if they cross him. She realizes that that's where she is at that moment, and the same again when Lex shows her Mason's body."

CHLOE: How do I know you're not just playing me here?
LEX: You're right, Chloe. I could be. I often bring high school girls to the morgue to show them what happens when trained professionals cross my father.

At the start of the episode, Josh Kelley was seen performing his new song live. "We thought, why not put someone on stage?" music supervisor Madonna Wade-Reed recalls. "This was the single on his new record, and the one that did the best. It was a great opportunity for his label to put him out there. They brought a playback tape and he sang along to it. The label flew him and his band out. We could have just had Lana and Chloe at the carnival, with carnival ride music, but it was nice to have the introduction to the mystery of the week in the middle of the music at the concert." ■

SMALLVILLE MUSIC

'Hey Now' by Black Toast Music
'Trouble' by Bonnie McKee
'Stupid Girl' by Cold
'Amazing' by Josh Kelley
'Over You' by Michelle Featherstone
'Trouble' by Pink

Smallville Ledger

FROM COVER GIRL TO JAILBIRD

Wayward Teen Repents

Lana Lang, the sweet fairy princess who graced the cover of *Time* magazine back in 1989, was arrested yesterday for stealing a car. Her boyfriend and crime partner, Seth Nelson, remains comatose in the Smallville Medical Center. Nelson was later the victim of an unrelated freak electrical accident. Doctors do not know if he will recover.

Yesterday afternoon, Nelson allegedly stole a new car off the Smallville Auto Sales lot on Highway 19. Nelson stopped by the Talon and picked Lang up.

It is believed that on Monday night, the duo broke into the county fairgrounds and rode the Ferris wheel. Neighbors remember seeing the ride lit up and running even though the attraction was closed. Sheriff Adams caught up with Lang on the outskirts of town and arrested the teen for her involvement with the stolen auto. Shortly thereafter, with Nelson's help, Lang escaped from jail.

Three days ago, before the electrical accident, Seth Nelson was working at his booth at the county fair and was rushed to the Smallville Medical Center after an altercation with a disgruntled fairgoer. Radiology department staff members reported that an incident occurred with an MRI machine while Nelson was being scanned but refused to further elaborate. The event is being investigated by the MRI manufacturer, as well as the center's administration.

After Nelson's accident, a repentant Lang turned herself over to the sheriff.

By Kathy Romita

SHATTERED

WRITTEN BY: Ken Biller
DIRECTED BY: Ken Biller

GUEST STARS: Kwesi Ameyaw (Darius), Patrick Bergin (Morgan Edge), Lorena Gale (Dr. Claire Foster), Caroline Marcelle (Young Chloe Sullivan), Angelo Renai (Surgeon), Darryl Scheelar (Jameson), Michael Daingerfield (Security Guard)

DID YOU KNOW?

During filming, Michael Rosenbaum caught someone trying to steal his van, and channeled his real-life frustration into his performance in his final scene with John Glover.

Lex tracks down Morgan Edge, who admits that he and Lionel were involved in the arson attack that killed Lex's grandparents. Lex is attacked by an assassin and flees to the Kent farm. However, there's no sign of a struggle at the mansion. Lionel continues to look for Lex, who he claims has had a nervous breakdown. Lex, meanwhile, is singing to a blanket that he believes contains his dead brother, Julian.

Lex becomes increasingly unstable as his story falls apart. Clark hides Lex in Lana's stables, and asks Lana to watch him. Lex has a paranoid fit and pushes Lana into a horse stall, seriously injuring her. He then confronts Lionel, who claims that the arson attack was Edge's idea, which sends Lex after Edge. Clark and Chloe deduce that Lex has been drugged. Clark also finds the assassin, and learns Edge's location.

Clark prevents Lex from shooting Edge, but the crime-lord uses kryptonite to attack Clark. Lex succeeds in wounding Edge and, in a moment of madness, attempts to shoot Clark, too. The wounded Edge tries to escape in his car, only to be stopped by Clark, who splits the vehicle in two in plain sight of Lex.

Lionel's security personnel arrive and take Lex off to an asylum, where his ravings about Clark are ignored. Lana tells Clark that she realizes that he's too dangerous to be around.

CHLOE: You know I want to believe that this is a conspiracy just as much as you do, believe me, but I think we need to start preparing ourselves for the possibility that Lex really is losing his mind.

"'Shattered' is a fantastic episode from top to bottom," Al Gough says firmly. "It's one of the best of the series, and Michael's performance is terrific. Everything just clicked. Ken Biller did a great job writing and directing. It was an episode where Clark lost — which is not something you often see with Superman, or on *Smallville*. There's no win for him on any front: Lex is in the asylum, Lana is hurt and tells him to stay away, so Clark is left alone."

The original plan had been for Rutger Hauer to reprise his role as Morgan Edge. "Unfortunately, he had a family conflict in Europe and couldn't do it, so that's why we brought in Patrick Bergin as Edge after plastic surgery," Gough explains. "The nice thing is that Rutger and Patrick are friends. We wanted someone who had the same build, so if you looked at him, you could believe that if they did plastic surgery, they could be the same guy underneath. Ultimately, if you're going to the plastic surgery 'well', which you can only really do once, you had better get the right guy. You don't

Opposite: Lex reaches his final destination.

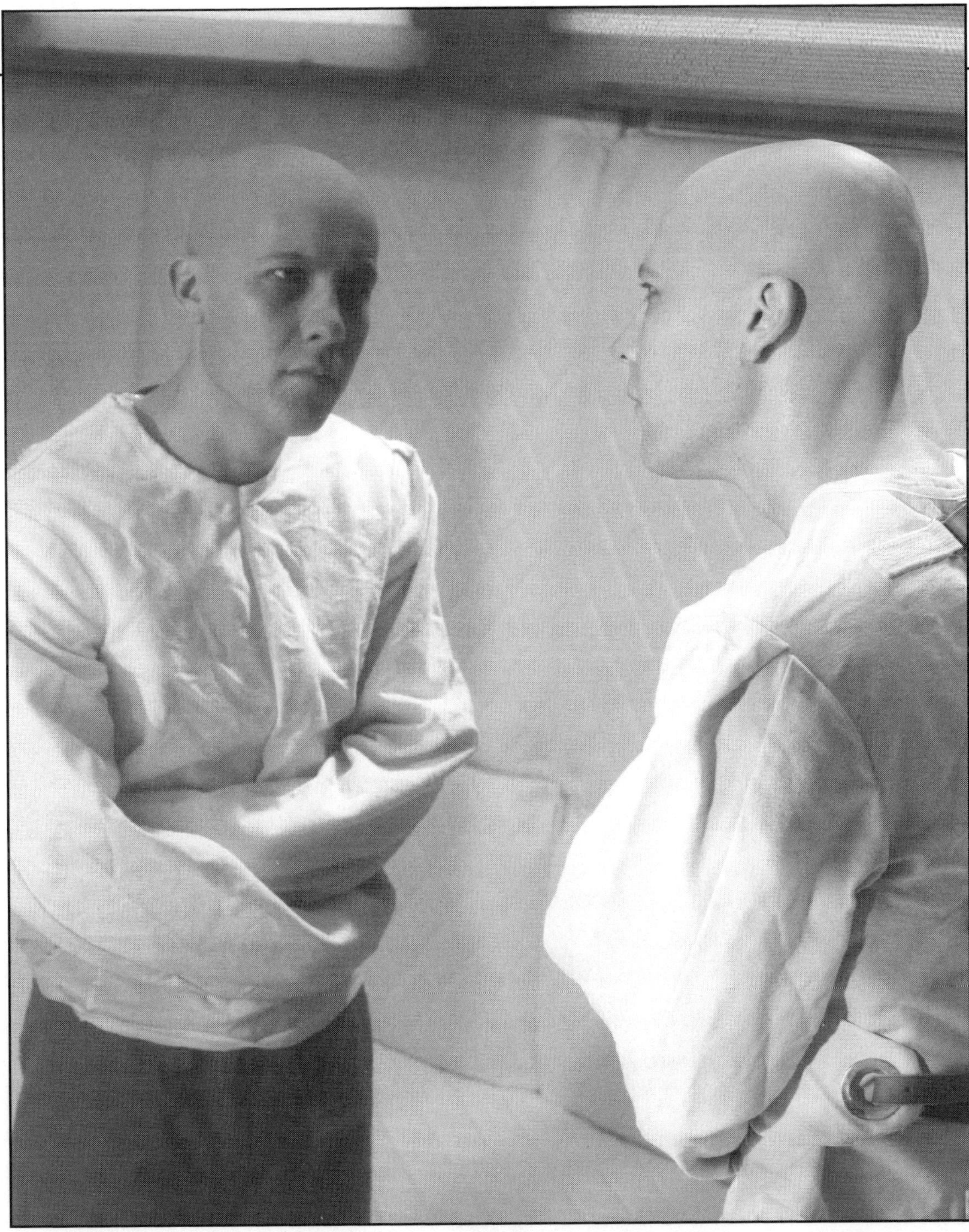

SMALLVILLE MUSIC

'Hurt' by Johnny Cash

want to have Rutger Hauer and then replace him with someone who doesn't look like him or isn't in the same league. When we got Patrick, we were pretty happy. But Edge is dead now — even in *Smallville*, a bullet to the chest means you're dead."

"Plastic surgery worked for this story," Mark Verheiden adds. "If you've got a billionaire who wants you dead, you probably would think about plastic surgery, so you don't look like the guy he's after anymore! The rest of the show is so great. We wanted to do a *Gaslight* episode, where we make it seem as if Lex is being driven crazy."

"That's why it was so interesting," Michael Rosenbaum notes. "It was a character study of Lex Luthor, and who he had become, and also of a man being driven into madness. It was a journey, and a great episode for the character, and me."

"I really enjoyed that one," Tom Welling says. "It was different, because Clark was trying to save his friend, and it meant a lot. The stakes were really high. Sometimes the person who's being saved or I'm tracking down is someone I don't know so much, but this was very close to home."

"I had to really sink my teeth into this episode," Michael Rosenbaum adds. "It wasn't something you could just 'walk through' as an actor. I didn't even sleep at night — I really dove into this role. Ken Biller and I worked well together. I really felt Lex couldn't just get crazy: he had to start out by thinking he was sane, then start wondering what was

Below: Lex tries to force the truth out of Lionel.

Smallville Ledger

LEX LOONEY?

Following is my account of the disquieting, gray morning when I visited Belle Reve at the request of my editor.

As I approach the asylum's impenetrable iron gates, I imagine its white-coated founding fathers chuckling at their cruelly ironic name: Belle Reve. Beautiful Dream. My guide into the abyss is one Wendell Connelly, a long-time Belle Reve orderly whose hunched, wary gait and emotionless face speak volumes about the depths of inhumanity he must have witnessed since he began working at the institution in 1992. We pass harried doctors and nurses with clipboards as our footfalls echo in the sparkling, antiseptic corridor leading to the staff lounge. We're not close enough to the inmates' doors to peer through the thick observation windows, but the padded, white walls inside are plainly visible, and the low, animalistic moans from some of the cells are impossible to ignore. Prominently placed cameras watch our every move.

We cross through the sterile, cheerless rec room, where dull-eyed patients sit blankly watching incongruously violent 1950s cartoons. Here, I get my first chance to observe Belle Reve's unfortunate guests up close. Never did I imagine that the disconcerting world so vividly portrayed in the motion picture *One Flew Over the Cuckoo's Nest* could still exist today.

By Christopher James Beppo

going on and if he was being framed, and then end up not even being able to think straight. Ken and I worked out an arc for the character, and because we don't shoot scenes in order, I had to trust Ken to maintain the continuity of the character and the through-line of the action that we had worked out."

LEX: Tell me, Clark! What's the price of a Judas kiss these days?

For Kristin Kreuk, the scene in the stable produced real bruises. "We were in the horse barn and there was a real horse in the stall at first," she recalls, "and then we decided that we wanted to get really close to Lana, so I got flat on my front, and Chris, our stunt coordinator, took two fake horse hooves and proceeded to stomp on me with them. The magic of the movies!"

Al Gough specifically wanted the Johnny Cash song at the end of the episode. "When this story was first pitched, I said I knew what song we wanted," he recalls.

"In our process of trying to clear the song and make our choices, Johnny Cash passed away," music supervisor Madonna Wade-Reed adds. "His heirs and the people who controlled the estate thought it would be a really nice way to honor him." ■

DID YOU KNOW?

Lorena Gale, who plays Dr. Foster, is also known for her role as the religious leader Elosha on the reimagining of *Battlestar Galactica*, another show which shoots in Vancouver.

ASYLUM

WRITTEN BY: Todd Slavkin & Darren Swimmer **DIRECTED BY:** Greg Beeman	**GUEST STARS:** Shawn Ashmore (Eric Summers), Lorena Gale (Dr. Claire Foster), Jesse Metcalfe (Van McNulty), Jonathan Taylor Thomas (Ian Randall), Ron Selmour (Connelly), Ian Somerhalder (Adam Knight), R. David Stephens (Doctor), Elias Toufexis (Luke), Gina Doty (Physical Therapist)

DID YOU KNOW?

Lana's physical therapy scenes with Adam Knight were filmed at one of Vancouver's many branches of the YMCA, which was dressed as Smallville Medical Center.

Lex is trapped inside Belle Reve Sanitarium, alongside numerous kryptonite-enhanced freaks, including Eric Summers (from the episode 'Leech') and Ian Randall (first seen in 'Dichotic'). Eric, Ian, and freak-hunter Van McNulty team up to attack Clark. McNulty bribes a guard to bring in a meteor rock, but Randall uses it to create a duplicate of himself, and kills McNulty.

Lionel forces Dr. Foster to use shock treatment on Lex to destroy his memory, specifically the information about his grandparents' death. Clark learns that it will happen within twenty-four hours, and tries to persuade Dr. Foster to cancel the treatment. She does, but is then killed in a car accident. Clark breaks into Belle Reve, but is attacked with the meteor rock and taken to the generator room where Eric leeches his powers once again. Eric and Ian fight, but Clark uses the meteor rock to stop Eric.

Despite begging his father not to proceed, Lex is given the shock treatment, which not only erases the memory of his grandparents' death, but also of Clark's powers — something Lionel only realizes when he sees a tape of Lex telling Clark that he knows his secret.

Lana's recovery is helped by another patient, Adam Knight, who claims he was injured trying to save his parents from a fire.

VAN: How do I know I can trust you freaks?
IAN: The enemy of your enemy is your friend.

"There's so much going on in this episode," Al Gough points out. "You've got the second half of the 'getting Lex out of the asylum' story, and we answer the question, 'Where do the krypto-freaks go when they leave Smallville?' And this is where they go — Belle Reve. It's a hospital in Superman lore which we've made into our hospital."

"I was hoping that they'd keep me in the asylum," Michael Rosenbaum admits. "The next three or four episodes would be about me trying to prove my innocence, and making more of an adventure of it. That would have been tough: being in a straitjacket, and going crazy and fighting every day wasn't the easiest thing to do."

The idea of various super-villains teaming up to fight the super hero is a staple of comic books, and it was an idea whose time had come on *Smallville*. "It was nice to see three krypto-freaks together," Al Gough says. "It was like the Legion of Doom getting together to take someone down. They all worked together to take down Superboy, and it really foiled Clark's plans. It was good to get Jonathan Taylor Thomas, Jesse Metcalfe,

Opposite: Lionel's unique brand of parenting in action.

DID YOU KNOW?

Over fifty pairs of the Belle Reve pajamas had to be prepared by the costume department for the various extras playing the asylum inmates.

and Sean Ashmore — we had a good popcorn component to the episode, as well as a good emotional element."

Bringing back Jonathan Taylor Thomas' psychotic dichotic Ian Randall allowed Mat Beck and his team a chance to improve the effects of his twinning. "We did the strands in 3D," Mat notes. "It gave us a little bit more control, and we could make them gooier and grosser."

Smallville has tackled the notion that Clark doesn't always win in a number of episodes, most notably season two's 'Ryan', in which Clark had to come to terms with the fact that he couldn't save his friend. "He's not God," producer Mark Verheiden notes. "He has limitations on what he can do. Even Superboy can't do everything. He did a lot in this episode, but he was incapable of saving Lex, who at that moment was probably one of his best friends, from this terrible thing his father was doing. It was an effective moment of tragedy on the show."

Al Gough notes that Lionel is taking these extreme actions out of a twisted love. "He was deliberately making his son mad, knowing that Lex had had this incident on the island," he says. "He just kept picking away at the crack in the foundations. He was definitely responsible — he loved Lex, and this was how he could protect himself and Lex, by wiping out his memory."

John Glover was taken aback when he read the script and saw the depths to which Lionel

Smallville Ledger

LEX LUTHOR BACK HOME

After a month's stay at Belle Reve Sanitarium, Lex Luthor has come home rested, relaxed, and raring to get back to work. "Lex has had an extremely stressful year," said Mitchell Taylor, spokesman for LuthorCorp and the Luthor family. "He just needed a break to sort out some issues."

The past year was indeed an emotional rollercoaster for the young tycoon. First, there was his wedding to Helen Bryce, followed by a plane crash that resulted in Lex being stranded on a deserted island for three months. Then, after Lex's dramatic return, tragedy struck again when the couple tried to resume their honeymoon, leading to another accident that left Lex's bride missing. Top it all off with a merger between LexCorp and LuthorCorp that gave Lex a major executive role in the conglomerate run by his father, Lionel Luthor, and it's easy to see how the pressure could build up and exhaustion could set in.

In particular, Lex's lonely ordeal on the island may have been more traumatic than doctors originally suspected. Following the caring advice of psychiatrist Dr. Claire Foster, Lionel Luthor agreed to admit Lex to Belle Reve.

Now back at Luthor Mansion, Lex appears completely rejuvenated by his time away, according to sources. He is expected to resume his day-to-day duties soon, supervising his father's diverse business interests. "Everyone at LuthorCorp Plaza is excited by Lex's return," Taylor noted. "The atmosphere here is positively electric."

By Frank Moore

SMALLVILLE MUSIC

'Future Proof' by Massive Attack
'So Far Away' by Staind

would sink. "I remember thinking that this season had really started to get pretty heavy for Lionel," he says. "They weren't fooling around anymore. Beforehand, I was testing Lex's strength, but this was heavy stuff." "Lionel's never been Father of the Year," Mark Verheiden points out. "But in 'Asylum', you saw how dark he was, and how awful he could be."

Above: Clark uses kryptonite to stop Eric.

JONATHAN: Someday, somebody is gonna make Lionel pay for his crime, but that day is not today, Clark. Now, I know this is very difficult for you, but you have to realize that there are people out there who are more powerful than you are.

Michael Rosenbaum thinks that 'Asylum' makes a good partner to 'Shattered'. "It was a fun episode for the audience, because we brought back a lot of characters, but it had a lot of character stuff, because we had to get one of our main characters out of trouble. It dealt with fun and friendship. Clark really wanted to help his friend, but at the same time, what would have happened if Lex *hadn't* had the shock? He would have remembered that Clark was hit with the car in 'Shattered'. The whole thing with the therapy was interesting, because the audience didn't know whether it would work, and how much he would remember."

Lorena Gale was disappointed that Dr. Foster was killed off. "I was a good guy, I think, just caught in Lionel's web," she notes. "But I would like to think that the time I spent with Lex, I was actually doing my job and helping him. Of course, he won't remember it now after the electroshock therapy." ■

WHISPER

WRITTEN BY: Ken Horton
DIRECTED BY: Tom Wright

GUEST STARS: Micah Alberti (Nathan Dean), Felecia Bell-Schafer (Judge Abigail Ross), William MacDonald (Masterson), Camille Mitchell (Sheriff Nancy Adams), French Tickner (Jewelry Store Owner)

DID YOU KNOW?

Actress Felicia Bell-Schafer (Judge Ross) played Jennifer Sisko, the ill-fated wife of Commander Benjamin Sisko on *Star Trek: Deep Space Nine.*

While trying to foil a jewelry store robbery, Clark is blinded when a kryptonite pendant reflects his heat vision back at his eyes. Conversely, his hearing becomes incredibly amplified. He hears Lionel admit to Chloe that Lex's job is a sham, and learns that Chloe has been spying on him. Clark tells Lex and Chloe what he's discovered, leaving Chloe in tears.

Nathan, one of the jewelry store robbers, kidnaps Pete. Judge Ross, Pete's mother, is told to dismiss the charges against the other robber or Pete will die. Clark uses his new power to locate Pete. Nathan uses a blowtorch against Clark before fleeing with Pete, but only burns off the scar tissue from Clark's eyes, restoring his vision, albeit blurred. Clark overhears a conversation between the two robbers, which enables him to locate Pete once more, just as his friend is about to be crushed in a car compactor.

Lionel cancels Chloe's column at the *Daily Planet*. She threatens to reveal his interest in Clark, and Lionel warns her not to make an enemy of him. Lex fires Chloe's father on Lionel's instructions.

Lana admits to Clark that she is interested in Adam, but doesn't realize that he can hear her tears as she drives away from the farm after telling him.

CHLOE: You know, I thought that you out of all people would understand making a stupid decision in a moment of weakness. But I guess I was wrong.

Every season of *Smallville* has introduced new powers for Clark Kent, as he slowly develops all the gifts that he will eventually use as Superman. After various hints were dropped that a new power would be introduced partway through the season, 'Whisper' finally revealed Clark's superhearing.

"We were going to do this earlier in the year," Al Gough recalls, "but the network told us that Tarzan had superhearing. We decided that whatever was done on other shows, we'd do better anyway." The update of Tarzan was short-lived and was off the air by the time that 'Whisper' was first shown, leaving Clark as the only person with phenomenal abilities in the auditory arena.

The episode also introduced another facet of Clark's future life — the glasses that will become an integral part of his disguise in Metropolis. "We did want to play the gag of Clark with glasses, and this seemed the way to do it," Al Gough admits. "Tom looked really good in those glasses," Kristin Kreuk comments.

Opposite: Mild mannered reporter-to-be Clark Kent.

To prepare to play blind, Tom spoke briefly with John Glover, who had portrayed

DID YOU KNOW?

In early drafts of the script, the robbers were going to be father and son working together, rather than the not-very-bright buddies that they became.

Lionel as blind during the early part of the second year. "I was concentrating on listening, on what was coming in through my ear, rather than through my eyes," Welling says. "As basic as that is, it's what I tried to do more than anything else."

Mat Beck admits that the depiction of superhearing will probably be tweaked as it goes along. "We want to zoom inside Clark's ear, and see the drum and the stirrup, and the hammer and all those things," he says. "It's difficult to do, but not impossible. When you talk about someone flying, it's something we all know, but this is superhearing, and we need to see a visual for the hearing, which needs to tie in."

Al Gough thinks that the episode ended up with "a lot of potentially good ingredients which didn't quite make a great cake. He shouldn't have been blind *and* got superhearing — Clark being blind could have been an episode to itself."

In fact, the original concept of the story saw Clark blind until the end of the episode. "It was determined that that wasn't going to be too satisfying to watch, so we gave him his sight back halfway through," Mark Verheiden recalls. "The rescue of Pete could have been more challenging. When he was blind, and he was forced to save Pete only using his hearing, that was more challenging, but once he got his sight back, it just became 'Let's find Pete.'"

Opposite: Lana asks Lex for advice.

Smallville Ledger

JUDGE'S SON SURVIVES KIDNAPPING

Pete Ross, the teenaged son of Lowell County Judge Abigail Ross and her husband, Bill, was kidnapped two days ago and managed to escape unharmed yesterday. Ross was abducted on the outskirts of Jonathan and Martha Kent's farm. Several hours later, Judge Ross received a phone call from a man threatening to kill her son if she didn't follow his instructions. She was ordered not to contact law enforcement officials and instructed to grant a motion to dismiss charges against Walt Masterson.

Walt Masterson was apprehended at the Smallville Jewelers after a botched armed robbery attempt a few days ago. Initially, authorities believed Masterson was the lone robber, but Ross' kidnapping has provided another suspect. Most of the customers in the jewelry store recognized the kidnapper, eighteen-year-old Nathan Dean, as being present during the robbery.

Dean was found unconscious in a heap of cars at an auto junkyard located approximately one mile from the airport. After kidnapping Ross and waiting for the judge to release Masterson, Dean planned to kill the Smallville High student. How Ross managed to free himself from the trunk of a car that was about to be crushed remains a mystery. After disabling the car compactor, he then called authorities, who found Dean at the scene. Sheriff Nancy Adams said, "This cat [Ross] just used up one his nine lives. What a remarkable kid. Quite a hero. He caught his own kidnapper and one of the jewelry store robbers too. I'm tickled pink he's alive."

By Kathy Romita

The major revelation of the episode was Clark's discovery that Chloe had been working for Lionel Luthor. "We weren't going to sell our series lead down the river," Al Gough points out. "Chloe made her deal, regretted it, and when Clark discovered it, she came clean and spent the rest of the season trying to bring Lionel down."

LEX: My father won a battle today.
CLARK: And Chloe's father is a casualty.
LEX: True. But the war's just starting, and I have the greatest advantage. He thinks I'm weak.

"It's interesting that because the audience likes Chloe so much, and how Allison plays her, to see Clark angry with her because she's colluding with Lionel is almost painful," Verheiden says. "You really want them to work it out. One of the reasons we wanted to have her collude with Lionel was to have something between her and Clark that was more than just Lana. When they were able to patch it up, it was a really nice moment." ■

SMALLVILLE MUSIC

'Cold and Empty' by Kid Rock

DELETE

WRITTEN BY: Kelly Souders & Brian Peterson **DIRECTED BY:** Pat Williams	**GUEST STARS:** Martin Cummins (Dr. Garner), Jim Thorburn (Max Taylor), Missy Peregrym (Molly Griggs), Suzanne Ristic (Max's Secretary, Maureen), Camille Mitchell (Sheriff Nancy Adams), Ian Somerhalder (Adam Knight), Cornell Turner (Security Guard), Chris Gauthier (LuthorCorp Technician)

DID YOU KNOW?

The character of Max Taylor was created for the Verizon commercials that appeared before the third season began on television. Jim Thorburn was invited to reprise his role.

Chloe asks *Daily Planet* reporter Max Taylor if he can publish an article she wrote about the Summerholt Neurological Institute. At school, she discovers that Lionel has taken back all the computers he gave the *Torch*. As she leaves, someone tries to run her over; it's Clark, who remembers reading an email and then no more. Later, Lana gets an email on her PDA from 'brainwave', which makes her attack Chloe. Adam manages to stop her, but Lana has no memory of the attack.

Max Taylor investigates Chloe's article and questions Dr. Garner. Molly, one of the former patients at Summerholt, promises Garner that she will stop Taylor from interfering — sending his secretary an email from 'brainwave' that makes her kill him. Lex hides Chloe at the mansion while Clark visits Summerholt. Clark learns where Molly is hiding, but is surprised to find Lex there. They realize that Molly was acting independently to try to stop Chloe from exposing Dr. Garner. Clark prevents his parents from attacking Chloe, while Lex takes Molly for safekeeping out of Garner's hands and returns the computers to the *Torch*. Clark and Chloe make up, and Lex blackmails Dr. Garner into helping him recover the memories he lost in the shock treatment.

Adam is evasive about his past, and is planning to leave Smallville, but Lana persuades him to stay and rent the apartment above the Talon.

LANA: Clark will always be a very important part of my life. But whatever there was between us is gone.

"'Delete' was based around an out-there idea," Al Gough admits. "Ultimately it's a really fun episode, just the idea that you can be gotten at through PDAs. It's similar to the movie *Telefon*, and also plays on the idea, which I think all good horror films do, that bad things happen to people at their weakest moments. The whole episode is Chloe's penance for working with Lionel on Clark, and then you learn that it doesn't have anything to do with Lionel at all, although Chloe still felt that kind of guilt."

The highlight of the episode for much of the audience was the knock down fight between Chloe and Lana in the Smallville High corridors. "That was an interesting script," Kristin Kreuk says. "I wasn't looking forward to doing the action sequence because it started off as two girls in the hallway beating each other up, and ended up in the showers in the girls' locker room. But I was promised over and over again that it

Opposite: Clark and Chloe examine the email evidence.

DID YOU KNOW?

Martin Cummins, who plays Dr. Garner, was one of the actors who auditioned for the part of Lex Luthor.

wouldn't be gratuitous in any way. It wouldn't be two girls having a catfight. And it wasn't. It was whole days of Allison and me, and our stunt doubles, fighting all over the school. I think it turned out really well."

Allison Mack shared Kristin's reservations. "It's always fun to know that I'm going to be working a lot, but I had concerns about the fight scene," she says candidly. "I had a few struggles, but it ended up looking really great. Our stuntwomen were so amazing. The fight scene took two days to shoot, and a day of rehearsal — that's two full twelve-hour days of shooting. It was the biggest fight scene we've ever had on *Smallville*, and Kim Chang and Lorelei Connelly just made us look so good."

Kristin and Allison didn't do any of the falls, but did pretty much everything else. "They would have us go through and do the fight scene up to the big falls, and then shoot our faces," Allison explains. "They would come in close so you could tell it was us. Then they would have Lorelei and Kim jump in and do the same things, but from a further distance, and go through all the way to the big falls and the punches. Then I would fight with Kim, and they would get all of my side, then I would flip Kim over, and they would do the 'cowboy change', where everyone freezes, Kim stands up and leaves, and Kristin comes in and we start fighting in the scene." Kristin adds, "In a lot of ways, it was one of the only true fight scenes in the show. Clark has all these powers — if you can throw somebody fifty yards, it's sort of the end of a fight!"

Below: Adam helps Chloe and Lana chase the trail through cyberspace.

"Both Kristin and I are physical women," Allison comments. "I train with all the stunt people on my time off for my workout, as does Kristin, and she has been doing karate all her life. We're both very coordinated and physical, and we can handle what we're doing in these stunts. We know our bodies well enough to be able to do it. I know our really wonderful stunt coordinators would never put us in any situation where we could actually get hurt. I don't hesitate to do anything they ask me to do."

SMALLVILLE MUSIC

'The Reason' by Hoobastank
'Try' by Nelly Furtado
'So Damn Lucky' by Dave Mathews

CLARK: You're not the same person you used to be.
CHLOE: Kind of like if I moved to Metropolis and became a motorcycle-riding party animal and told you to get out of my life. Kind of like that, right?

'Delete' also introduced some ideas that would resurface later in the season, as well as giving the first mention on *Smallville* of Chloe's cousin, and Clark's future wife, Lois Lane. "We reintroduced the idea of Dr. Garner and Summerholt, and Lex trying to get his memory back," Al Gough notes. "That would then be reintroduced in 'Memoria'. It was an out-there premise, but when you get to the end of the episode, it was fun." ■

Smallville Ledger

URBAN LEGEND: ABLE TO LEAP BUILDINGS IN SINGLE BOUND?

This is the third in a series of four articles that chronicled the exploits of Metropolis' Urban Legend this summer. It is reprinted with the permission of our sister paper, The Daily Planet.

Metropolis is an island steeped in legend and lore. As a reporter who occasionally loves to break into columnist mode in order to bring the more fascinating tales of this city to my readers, I am one part baffled and two parts intrigued by the most recent subject that has every one of you abuzz at our local coffeehouses, bars and water coolers. You know the question; it barely needs repeating. But here Max blows: Who in Hell's Kitchen is the Urban Legend? I mean, come on people! Is this guy for real?

No one really knows the answer to that question, yet the folklore continues to flow. My contacts at the MPD remain baffled as they press on, interrogating the few who claim to have actually seen him (her? it?).

Are we talking about a real human being here? Or is this just another chapter in a growing Urban Legend the entire city is spinning together? For now, we are left without a concrete answer or a positive ID. For now, we are simply left to wonder, ponder, and discuss. For now, I'm still Max Taylor, and this is what you'll talk about today.

By Max Taylor

HEREAFTER

WRITTEN BY: Mark Verheiden & Drew Greenberg **DIRECTED BY:** Greg Beeman & James Marshall	**GUEST STARS:** Joseph Cross (Jordan Cross), Shaun Johnston (Mr. Cross), Greg Kean (Mr. Altman), Camille Mitchell (Sheriff Nancy Adams), Meghan Ory (Megan), Ian Somerhalder (Adam Knight), Louise Grant (Elderly Lana)

DID YOU KNOW?

According to reports on the Net, one fan was so surprised by the sight of the cape that he spilled a soda in his lap.

When new student Jordan Cross touches someone, he sees his or her death. However, when he touches Clark, he only sees a red cape with a yellow insignia. Clark saves Mr. Altman from committing suicide, but the coach wishes he hadn't been stopped, as he feels responsible for his daughter Julie's death. The coach overhears Lana and her friend Megan talking about Julie leaving a party after an argument with Megan.

Jordan touches Megan and sees her suffocating in a fire. Later, when he and Clark try to find her, she's vanished. He discovers that Lana is also in danger of the same fate.

Altman has kidnapped the girls, and plans to start a fire and burn himself to death alongside them. Clark and Jordan rescue Lana and Megan, but Altman dies in the fire. Jordan realizes that he is no longer getting visions.

Adam refuses to discuss his nightmares. Chloe spots him injecting himself and tells Lana. Lex warns Adam that he's investigating him, and Adam then ignores Lana when she asks him about drugs. Chloe finds an vial of liquid in Adam's apartment. Jordan tells Lana that when he touched Adam earlier, he realized that Adam had died in a fire some months earlier.

And Clark discovers his father having a heart attack in the barn...

CLARK: What'd you see?
JORDAN: Like you don't have an end like other people. It's like you live forever.

Executive producer Al Gough refers to 'Hereafter' as a kind of 'greatest hits', "The K-Tel of *Smallville*: things you've seen in a lot of other episodes, you get to see in this one episode." But there are also a couple of rather significant new elements: "You get to see the cape, and Jonathan has his heart attack."

"We wanted to explore an idea that was brought up in season one's 'Hourglass'," writer Mark Verheiden explains, "with Clark thinking that he might never die. There's a scene in that episode where he's looking at all the graves. We used this kid who sees the death moment in people to touch Clark and go, 'I don't see that for you.' We were exploring the idea that Clark is an alien, and being what he is, is different on an almost cosmic scale. He has no death, so that takes him outside the metaphysical reach of this kid's ability. It also enabled Clark to prevent the things from happening that the kid was seeing — because he was so outside that, he was able to save Lana and Chloe. We really wanted the kid to be very tortured — it's not a great power to have, to know when someone you know is going to die and you can't do anything about it."

Opposite: Clark rescues Jordan from the flames.

SMALLVILLE MUSIC

'100 Years' by Five for Fighting
'Two Steps Closer' by Static
'I Owe You' by Stegala Music
'Rhapsody on a Theme of Paganini, Op. 43' by Sergei Rachmaninoff

The moment when Jordan touches Clark and sees something he can't explain was one of the highlights of the season. "The cape came after a lot of discussion," Verheiden says. "We were trying to figure out the best way to show his cosmic lifespan. We thought of having Clark lying there with someone who looks like Lex dumping kryptonite on him and killing him. We spent quite a while mulling over the different possibilities, and I think it was Miles who came up with the idea of the cape. It's a great iconic moment — it's the first time in the show you ever saw any hint of the costume. If you're a Superman fan and really into the iconography of this stuff, you got chills down your spine."

LEX: Lana, I've had people assume the worst about me all my life. Trust me, it's hard to respond when people always jump to the wrong conclusions.

"That was done entirely in CGI," Mat Beck explains. "A lot of people had input into that, and John Hahn was the guy who pretty much put it together. We wanted to show the cape but not have it entirely clear what it is you're looking at. You get a hint of the 'S', but not quite. I wanted to start out with something that made you think, 'What the hell is that?' Then gradually you get enough information to realize it was the cape, but at the same time you're flying through the cosmos. For all I say

Below: Clark desperately calls for help after Jonathan's heart attack.

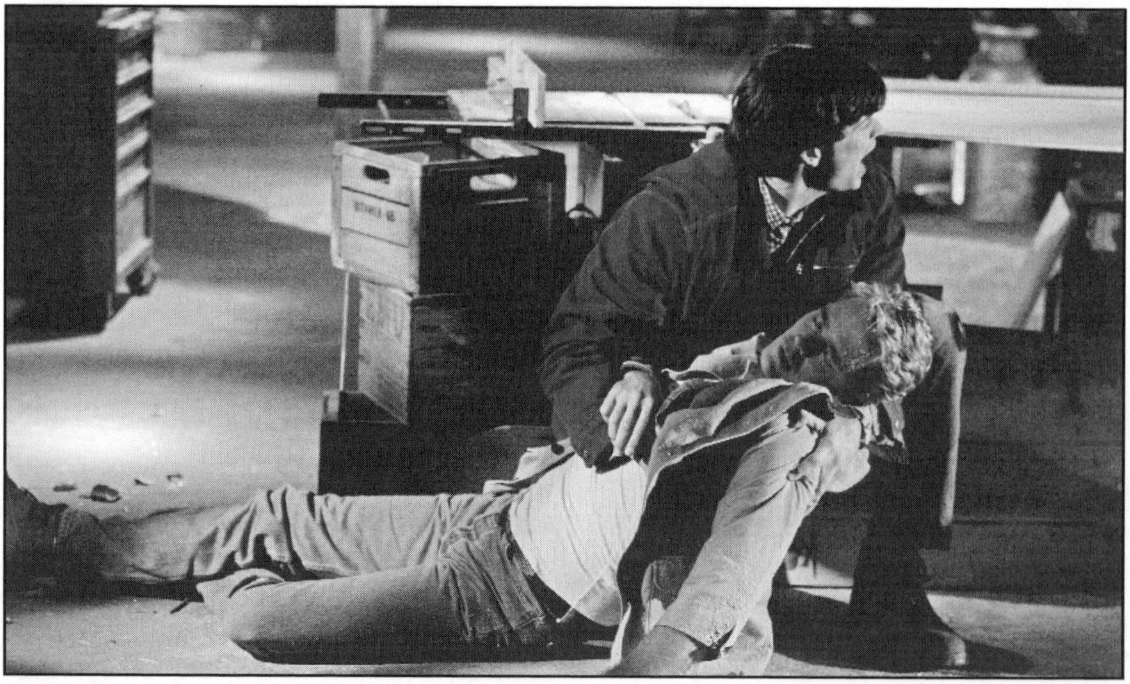

about shooting the live action stuff right so we can put it together in post-production, sometimes you can put a shot together without shooting a damn thing!"

Jonathan's heart attack might not have been a surprise to regular readers of the Superman comic books or fans familiar with the Christopher Reeve movies, but the producers were determined that there should be a reason for it within the continuity that they had established. "We were always trying to find reasons for the mythology," Miles Millar says. "The reason Jonathan has a heart condition is because he took the powers to bring Clark back. The sacrifice of the father is that he'll sacrifice anything to get Clark back — so he sacrifices his health."

"We had talked about it as early as season one," Mark Verheiden recalls. "We came very close to using it in season two — 'Fever' started off as the one where he'd have the heart attack. The idea that he made a deal with Jor-El meant it wasn't just a 'one day I was out on the tractor and I fell off' story. Giving him the heart attack was a really great way to amp up the stakes with Jonathan, and show how far this family would go to protect their son." ■

DID YOU KNOW?

James Marshall took over direction of 'Hereafter' after original director Greg Beeman became seriously ill partway through shooting.

SMALLVILLE TORCH

COACH ALTMAN DEAD, THREE STUDENTS RESCUED

A dark life ended darkly last night as a grieving man got his suicidal wish and almost took two innocent teenagers with him.

Three Smallville High students were saved moments before a huge explosion emanating from the basement of a house rocked Lowell County. The blast, which was felt as far away as Grandville, claimed the life of SHS cross-country coach Joseph Altman.

Just one month ago, Altman's brain-damaged daughter was taken off life-support. She had been in a coma for four years. Altman had attempted suicide earlier in the week but was saved when junior Clark Kent pushed him out of the way of an oncoming car.

Following the incident, Altman was placed under supervision at Smallville Medical Center, where he repeatedly insisted that he wanted to be with his daughter. Upon his release, Altman apparently remained depressed and delusional, blaming his daughter's death on Lana Lang and Megan Calder, who were cheerleaders with Julia Altman in junior high — a ludicrous accusation considering Julia was walking home alone when a car sideswiped her.

Altman abducted the two girls and attempted to set them on fire in the basement of his home before detonating the explosives. Clark Kent and new student Jordan Cross pulled Lang and Calder out of the house. Cross, suffering from smoke inhalation, was in turn saved by Kent. The Calder family and Lana Lang are not considering a lawsuit against Smallville Medical Center for releasing Altman.

"Coach Altman's death saddens us all. First Walt Arnold, then Mr. Kwan, and now this. It is a sad day for Smallville High," commented Principal Terrance Reynolds. Details of Altman's memorial service will be announced soon.

By Chloe Sullivan

VELOCITY

WRITTEN BY: Todd Slavkin & Darren Swimmer
DIRECTED BY: Jeannot Szwarc

GUEST STARS: Moneca Delain (Mara), Ryan Merriman (Jason Dante), Camille Mitchell (Sheriff Nancy Adams), Ian Somerhalder (Adam Knight), Françoise Yip (Dr. Lia Teng), Colin Corrigan (Mechanic)

DID YOU KNOW?

Director Jeannot Swarzc was used to handling the problems of superpowers after helming the 1985 big screen version of *Supergirl*.

Pete is getting a thrill from winning streetcar races for Jason Dante, who is using a kryptonite-enhanced fuel injection system. Pete refuses to throw a race and Dante tells him to find $20,000 or die. Pete asks Clark to steal the money. Clark refuses, but asks Lex for help. Lex refuses as well when he learns why. Clark steals Lex's Porsche so Pete can have one final race against Dante. Clark then overhears Dante's men discussing the bomb that will explode in Pete's car when he uses the kryptonite fuel to reach 100 mph, but Clark is hit with a cylinder of fuel and dumped in Pete's trunk, next to the kryptonite. Clark manages to punch through Pete's gas tank in time, but Dante's car flips wildly on the spilled fuel and crashes, killing Dante. Clark returns the Porsche — although Lex guesses what has happened — leaving his friendship with Pete severely damaged.

Lana confronts Adam, who admits he is really Chad Nash, who died five months earlier. Chloe tells Lex that the serum she found contains unearthly blood platelets, and that Dr. Teng from Metropolis University is tending Adam. Dr. Teng refuses to discuss Adam with Lex, so he has her funding withdrawn. He tells her she can have unlimited funding if she reveals the source of the platelets. Adam threatens Lana to stop her from telling anyone about him.

JONATHAN: As hard as this is for you to believe, not everything that goes wrong in Smallville is Clark Kent's fault.

"I never understood people's problems with 'Velocity'," executive producer Miles Millar states bluntly. "It's a completely legitimate episode of the show. We have to do different stories. To me, it's a really fun, high concept, cool, well-shot show. The interesting thing is that shows the fans hate, we often get our highest ratings for."

Al Gough agrees. "I really liked the episode, and I don't understand the gripe over it. Sam is terrific in it, but it proved to us once and for all that for all the grousing and comments online, people don't really care about Pete."

The idea of one of the characters becoming involved with drag car racing had been in the producers' minds since the first season, at a time when it was becoming a problem in the real world on the streets around Vancouver where the series shoots. "It's something that happens in small towns," Gough points out. "We wanted to do it with our people, and now that Pete knew the secret, we had a way to make it about one of our characters and tie Clark in very strongly. It's a very strong teen episode with a really good, high concept."

The emotional impetus behind the episode was the way in which Clark and Pete's

Opposite: Pete gets on the wrong side of Dante and pays the penalty.

DID YOU KNOW?

Ian Somerhalder was the winner of the 2002 Young Hollywood Award for Most Exciting New Male Face.

friendship had changed dramatically over the past few months. "We wanted to see Pete finally go, 'You're not my friend anymore, Clark," producer Mark Verheiden recalls. "'You're hanging around with Lana and Lex, so don't tell me that I can't go and do this with my new friends.' But then, when Pete's in over his head, he's forced to go to Clark and get him to steal some money from Lex. It was to show that these two friends had fractured a bit, and that sometimes there's no coming back. At the end of the episode, Pete tries to apologize, but Clark doesn't know if they can go back."

PETE: Clark, you're awesome!
CLARK: I'm glad you're happy, Pete. Because I had to lie for you and I had to steal for you. Now I have to cheat for you.

Early versions of the script saw the two starting to play hoops at the end of the episode, as if nothing had happened, but this was altered into the more downbeat conclusion seen onscreen. "I think that was ultimately better," Verheiden says.

It also marked the start of Pete's growing alienation from the other characters, which

Smallville Ledger

SMALLVILLE'S DUBIOUS DRIVING DISTINCTION

Street-Racing Fatality Underlines Town's Alarming Accident Rate

Representatives from the Kansas Highway Patrol (KHP) and the Kansas Department of Transportation (KDOT) joined Smallville Mothers Against Reckless Teens (SMART) to condemn the incursion of illegal street racing into the byways of Lowell County at this week's Women's Association meeting.

"The death of teenager Jason Dante [in a street-racing accident] this week is tragic enough. But just as disturbing is the underground culture that has been laid bare," stated SMART president Lorrie Ann Martin. "We are now learning of activities that have been going on right under our noses for many months, carried out in secret locations by a very well-organized club of unsupervised teenaged miscreants."

Patrol Superintendent Selwyn Thrush warned Smallvillians that they are in danger of coming into line with the rest of the country in terms of driving-related deaths. "People look around and think 'It could never happen here,'" he commented. "But this is not such a small town anymore. You're starting to see big-city problems. Hopefully this incident, and the one out by Saunder's Gorge last year, are not the tip of the iceberg for Smallville.

"Frankly, Smallville is already infamous among Kansas law enforcement because it well exceeds the state and national average of per capita automobile accidents, and some of those incidents are pretty far-out. The last thing you need here is people purposely driving recklessly."

By Jim Bradlee

Above: Lex and Clark spar verbally about the nature of friendship.

would culminate in his departure at the end of the season. "There was a building understanding in Pete that knowing Clark's secret was more than he could deal with," Verheiden comments. "We had thought that once he knew the secret, it would make Pete a much more integral part of the setup, and bring him into the stories more, but it didn't."

The music soundtrack was markedly different from the usual episodes of *Smallville*. "I had heard about [British hip-hop artist] Dizzee Rascal and was dying to use something," Madonna Wade-Reed says. "I got the record, went to license it, and turned out I was the first person in the United States to use it. It was a real task to use it. It was very much an 'out of the box' choice for the show, but they let me do it. The whole episode was very much like *The Fast and the Furious*, and it was also very focused on the character of Pete, so being the only black actor, we had the opportunity to use music we would otherwise never have used. If you try to use hip-hop sometimes, it would be like oil on water — but that music is very synonymous with what was going on. It had to set the people who were doing this racing apart from other townsfolk — it was a group of people you didn't know lived there. They got to sound really different, and from the music perspective, we got to really push the boundaries." ■

SMALLVILLE MUSIC

'Fix Up, Look Sharp' by Dizzee Rascal
'The Way I Am' by Knoc-Turn'Al
'You Know' by Saq
'Evilution' by Shocore
'Rawkfist' by ThousandFootKrutch
'Wonderwall' by Ryan Adams

OBSESSION

WRITTEN BY: Holly Harold
DIRECTED BY: James Marshall

GUEST STAR: Lynda Boyd (Mrs. Baker), Sarah Carter (Alicia), Camille Mitchell (Sheriff Nancy Adams), Paul Perri (Mr. Baker), Ian Somerhalder (Adam Knight), Françoise Yip (Dr. Lia Teng)

DID YOU KNOW?

Lynda Boyd and Sarah Carter, who play mother and daughter in 'Obsession', both appeared in the movie *Final Destination 2*, although they do not play related characters.

During a tour of LuthorCorp, Clark and classmate Alicia Baker are nearly killed in a faulty elevator. Clark uses his powers to smash the doors, but Alicia teleports them both out instead. Sharing their secrets, the two bond and go to the movies, and Alicia later teleports into bed with Clark just as Jonathan walks in. Alicia disappears, and Clark admits to his parents that she knows about his powers, but not his weakness — not realizing that Alicia has reappeared in the house, and is eavesdropping. Later, Alicia's parents admit to Clark that they have a lead-lined room from which she can't escape. Alicia is furious that they told Clark, and hospitalizes her father. She then puts her picture all around Clark's loft, and scratches Lana's picture out of Clark's yearbook.

With Chloe's aid, Clark tries to capture Alicia in a lead paint-lined room, but she uses a meteor rock to weaken Clark and goes after Lana. Clark covers the rock in lead paint and goes to Lana's rescue, covering Alicia in the paint to prevent her from teleporting away.

Lana finds Adam's journal with notes about her and Clark. She orders him to leave, but he refuses. She warns Clark about Adam's spying, then asks Lex for help. However, when Adam vanishes, Lex denies responsibility; in fact, Lionel has Adam incarcerated and tells him he is now withholding the special formula that Dr. Teng developed to keep Adam alive, as the youth failed in his task of spying on Clark...

CHLOE: Welcome to the Lionel Luthor exhibit. Next Floor: Victims, sycophants, and hatchet men.

"'Obsession' was a great episode, purely because you got to see a sexual chemistry between two of the characters. I don't know that you get to see that much on *Smallville*," Kristin Kreuk points out. "It's normally a romantic chemistry. That was really about sex, and revealing who you are."

Executive producer Al Gough also liked the episode a lot. "It's one of the few times where you bring in the guest star of the week to be the love interest and you really believe it," he notes. "Tom and Sarah had real chemistry. We tried to do it with Kyra, the wolf girl in season two's 'Skinwalker', but you didn't quite believe it. With this character you did."

The episode also features what Gough calls "the great male fantasy turned nightmare — the moment when your dad walks in." Producer Mark Verheiden adds, "It was a classic moment — caught in bed with a sexy girl. The actress was very hot, which made for great scenes anyway. Clark finally found someone he could bond with, because they

Opposite: Sparks fly between Clark and Alicia.

SMALLVILLE MUSIC

'Setting of the Sun' by Ben Jelen
'Echo' by Blake Hight
'One Thing' by Finger Eleven
'Parking' by The Fuzz

shared a secret of extraordinary abilities."

"You would completely believe that, were Clark not [in love] with Lana, he could give it a whirl with her," Al Gough continues. "That's why we decided to bring her back for season four."

LEX: You go on one date with this girl and already she's sneaking into your bedroom, putting sexy pictures of herself in your locker, and practically asking your parents to book the wedding chapel. What's your secret, Clark?

The sequence in which Clark and Chloe try to trap Alicia in a room lined with lead paint evolved from discussions that originally took place for a scene in season two's 'Visage', where Clark was trapped in the Kents' storm cellar by Tina Greer, who placed Lana's kryptonite necklace nearby to keep him powerless. "It was an idea I came up with for that moment in 'Visage'," Mark Verheiden says. "There would be some old lead paint in the cellar — they don't make it here in the United States any more, so it would have to be quite old — and Clark would then kick it over the kryptonite, which

Smallville Ledger

TEEN TURNS TO VIOLENCE
Daughter Suspected In Father's Fall

It appears that Wednesday night's drama at the Talon isn't quite over. In a surprising final act, Alicia Baker, who was arrested for her violent attack on Talon co-owner Lana Lang, is being questioned about a separate, yet equally vicious incident. Ms. Baker may be the prime suspect in her father's recent accident.

John Baker, 42, was rushed to Smallville Medical Center earlier this week for injuries sustained in a fall. While the senior Baker claims to have slipped on the basement steps, Sheriff Adams points out that his story has several inconsistencies. "There's a blunt trauma injury to the back of Baker's head that doesn't jive with any tumble I've ever seen," says Adams. "It looks to me like somebody took a good whack at him." While Sheriff Adams admits that John Baker has refused to point the finger at his daughter, or even acknowledge that there was an assailant, Adams remains confident that she and her deputies will soon have sufficient evidence to pin the crime on Alicia Baker.

According to the sheriff, Alicia Baker assaulted both her father and Lana Lang in a futile attempt to attract the attention of fellow Smallville High School student Clark Kent. "Mr. Kent spurned her affections," Adams surmised. "We've all seen *Swimfan*. We know what happens in these situations."

Adams had no comment about rumors that the basement where Mr. Baker's accident occurred was locked from the inside.

By Lacey Lee

DID YOU KNOW?

Internet rumor originally suspected that Adam Knight was really going to be a younger version of Dark Knight Bruce Wayne — but this was never one of the possible routes for the character.

would allow him to get out and save himself. It changed and became the ship saving him, but the paint thing came from there."

As ever, the crew were warmly welcomed when they returned to Templeton High School, which spends part of its year doubling as Smallville High. "The principal of the school used to be in the film business," executive producer Greg Beeman recalls, "so he knew a lot of people on our crew. He was happy for us to film there, but then we just started encroaching more and more. By the end of season one, we'd painted about half their school in our colors. There are giant Crows logos, and it says Smallville High everywhere. They've now even changed their school colors to match our colors! Our shooting days are during term time, so there's this bizarre interface that happens where we're in the middle of working in the hallways, and the school bell rings. Three hundred kids pile into the hall, we stand aside, they move our lights out of the way while they open their lockers, then close them, leave, and we get back to work. They're all really proud of the fact that their high school is Smallville High, but they're mainly oblivious to us — except when Tom Welling shows up, and then all the girls lean out of the windows going, 'We love you, Tom!'"

Above: Clark ponders life with Alicia.

Although it wasn't as obvious to American audiences during the show's broadcasts, overseas fans immediately noticed that the music over the closing titles for the third season was markedly different from the two previous years. "Everyone had called me up during the second year saying that the old music wasn't really what the show was about anymore," composer Mark Snow recalls. "It had been one of the numerous attempts to come up with a theme before they found the Remy Zero song. The new music is not quite as in-your-face heroic, it's melodic. It seems more an outgrowth of the generic *Smallville* score." ■

RESURRECTION

WRITTEN BY: Todd Slavkin & Darren Swimmer
DIRECTED BY: Terrance O'Hara

GUEST STAR: Julian Christopher (Dr. MacIntyre), James Kirk (Garrett), Camille Mitchell (Sheriff Nancy Adams), Tahmoh Penikett (Vince), Jerry Wasserman (Dr. Yaeger), Françoise Yip (Dr. Lia Teng), Sandra Ferens (Morgue Doctor), Bruce Harwood (Ben Paul/Medvac Pilot)

DID YOU KNOW?

James Kirk, who played Garrett, appeared alongside Kristin Kreuk on the Canadian high school drama *Edgemont*.

Jonathan reluctantly agrees to have a triple heart bypass operation. At the hospital, Clark meets his classmate Garrett Davis, whose brother Vince dies from liver disease while they are there. Martha offers to look after Garrett at the farm until his uncle arrives. However, Vince has been injected with a special formula by Dr. Teng, which revives him. He arrives at the Kent farm, claiming he escaped from Dr. Teng — but Vince collapses, and the doctors tell Garrett his brother will die without an immediate liver transplant. Lana tells Clark that Adam was using a similar-sounding drug to stay alive. Clark finds Dr. Teng's lab, where he overhears Lex and Teng discussing the formula, and he steals the vial.

Garrett arrives at the hospital with a kryptonite-enhanced bomb strapped to his chest, demanding Vince is given a transplant immediately. Clark persuades Garrett to release his hostages in exchange for the vial, but once it comes near the kryptonite Garrett is wearing, it becomes useless. Garrett demands that the doctor use Jonathan's liver. Sheriff Adams orders a sniper to shoot Garrett, and Clark gets the bombs out of the hospital before they explode. Jonathan's bypass is successful. Dr. Teng tells Lex that she is keeping Adam alive against Lionel's orders, but he is dying. And Clark realizes that the serum being used has been taken from his blood...

JONATHAN: Now, there's no doubt in my mind that you and Clark can take good care of each other, but you have to remember our promise. When he's ready, you have to let him go.
MARTHA: When that day comes, you'll be standing right beside me.

'Resurrection' is the episode in which the liver disease plotline that began with Adam Knight took center stage. When Knight was introduced in 'Asylum', the original intention had been that the character would be a new boyfriend for Lana. "It was supposed to be a legitimate relationship," executive producer Al Gough points out. "Lana had broken up with Clark, and she was going to have another boyfriend, but the nature of the actor and the chemistry wasn't working. It made us go into a more thriller *Pacific Heights*-like direction — you've got a tenant that you can't get rid of. What is he doing? Unfortunately, it then devolved into more sci-fi than we ever wanted to go. It was very generic, in a weird way. This is where we're in the middle of the secret labs portion of our season, and we knew that we had to get out of it as quickly as we could.

Opposite: Chloe and Lana in the *Torch* office.

DID YOU KNOW?

Julian Christopher made his final appearance to date as Dr. MacIntyre, having previously been seen at the SMC as far back as season one's 'Leech'.

Unfortunately with a TV show, you can get into a plotline that isn't working, and you then get out as soon as you can. The arc with Adam was six episodes that wanted to be one thing and didn't get there, which was frustrating."

"If you had to pick an episode that epitomized the scientific stuff, it would be 'Resurrection'," producer Mark Verheiden adds. "It felt like we were getting awfully medical in that episode. Jonathan's in the hospital waiting to get a heart operation, and there's a kid with a cloned liver trying to figure out how he was brought back from the dead in this secret lab with a bunch of monkeys in it!"

Early versions of the story saw Garrett bonding with Clark over Jonathan's illness because it was his father, rather than his brother, who was dying. "I think he was going to be a bit more sympathetic originally," Verheiden recalls. "It got darker as we went along, and he got shot and killed by a sniper. The villains sometimes don't matter on the show — they're secondary, engines to keep the other great stuff going — but all this kid wanted to do was help his brother, and we blew his brains out."

For Camille Matthews, 'Resurrection' showed a side to Sheriff Nancy Adams that she had not suspected was there. "I was surprised when I read the script," she admits. "I wasn't expecting it. I don't know why I was surprised, because she's obviously someone

Smallville Ledger

SURGERY DURING A SIEGE

While Saving Lives, Doctors Sometimes Risk Their Own

The shock waves will continue long after the mess in Smallville Medical Center's waiting area is cleaned up. Already the debate over unionizing doctors is being revived, and security measures are being beefed up — somewhat akin to fixing the barn door after the horses have escaped.

The circumstances call to mind a similar situation in 2002, when doctors at Chicago's County General Hospital staged a walkout to protest the lack of metal detectors in their emergency room. A chief resident led the revolt, two years after surviving a vicious workplace attack that also took the life of his female colleague.

Asked if doctors at SMC will have to resort to such tactics, Scanlan commented, "Granted, we're not a major urban hospital like County General. We don't have riots and gunmen in the emergency room every third week, or helicopters falling off the roof and Army tanks smashing into the ambulance bay. But there has to be a certain standard of protection even in Smallville. After Dr. Helen Bryce was attacked and robbed in her lab, we all thought nothing like that could ever happen here again. But here we are, less than a year later, performing delicate cardiac surgery while someone is getting ready to blow us all sky high. It's unreal."

By Jim Bradlee

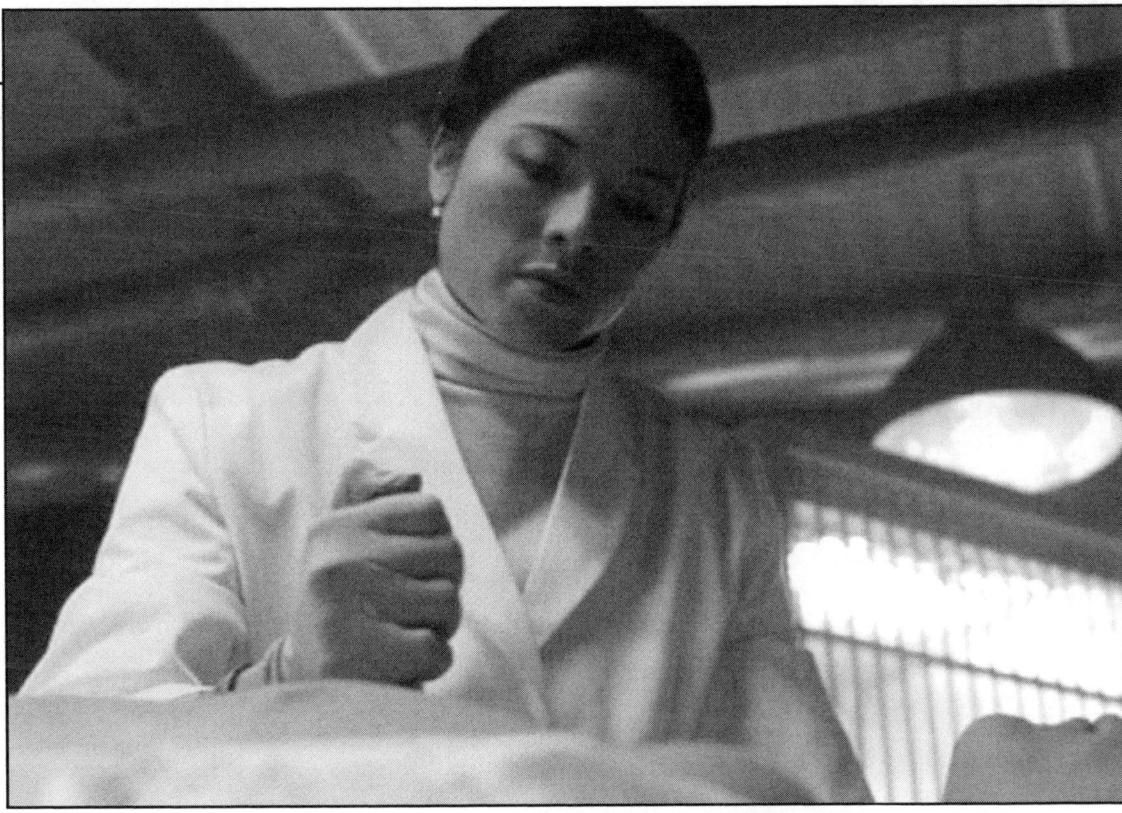

Above: Dr. Teng operates on Adam.

capable of doing whatever it takes to protect the community, but I was surprised that the character would finally go ahead and give the order to have someone killed. She doesn't have a choice in that necessarily, but I found it interesting that they chose to reveal that tough side to her. I think that, in terms of who and what she is, that was the most revealing and truthful moment."

CLARK: I think that Lionel Luthor has found a way to use my blood to bring people back to life. If my blood can save lives, maybe I should come forward. What if there's something inside me that can repair dad's heart permanently?

Madonna Wade-Reed was pleased with the choice of music to underline the moment between Lana and Lex. "The scene was kind of standing on the edge of the cliff with both the music and the acting, but they don't actually take the leap. The song ['Infatuation' by The Rapture] was very moody, and supported the mystery of 'Are we ever going to figure out what these two people think of each other?'"

Al Gough concludes that for him, the highlight of 'Resurrection' is simply the "cool sequence at the end of the episode where we go into Clark time, with the contrails." ■

SMALLVILLE MUSIC

'Everything' by Alanis Morrisette
'Infatuation' by The Rapture

CRISIS

WRITTEN BY: Kelly Souders & Brian Peterson
DIRECTED BY: Ken Biller

GUEST STARS: Cedric De Souza (Dr. Burgeon), Camille Mitchell (Sheriff Nancy Adams), Johannah Newmarch (Gwen Burton), Aaron Pearl (Glenn Burton), Ian Somerhalder (Adam Knight), Françoise Yip (Dr. Lia Teng), Terry Chen (Detective Paul Kabe), Deryl Hayes (Second Detective), Gary Hudson (FBI Agent)

DID YOU KNOW?

John Glover was informed that Lionel was suffering from terminal liver disease over lunch in New York, during the filming of his scenes with Christopher Reeve for 'Legacy'.

At the Youth Center, Clark picks up a call on the crisis hotline from Lana, who is screaming for Clark's help as she is attacked by Adam. But Lana is actually at the Talon, perfectly fine. Clark, Lana, and Chloe deduce that somehow, the call has come back through time.

Dr. Teng gives Adam one last dose of medicine, which gives him twelve hours to live. Adam attacks the doctor and escapes. Lex takes Clark to the secret lab, where they find Teng and several others dead, and Adam gone. Adam kidnaps Chloe and uses her to capture Lana, but Lana manages to break loose from him to make the call. Lightning hits meteor rocks by the telephone pole near the Youth Center, sending the voice back into the past, but Clark manages to catch the end of the call, and Lana tells him they're at a storage facility. Clark superspeeds in front of the bullet that Adam fires at Lana, deflecting it. Adam then dies for the last time.

Lex agrees to help the FBI bring down his father in exchange for immunity over Adam's crimes. Lionel briefs a new doctor to replace Dr. Teng, then reads a report stating that he has terminal liver disease. He takes a handgun out of its box and places the barrel in his mouth…

CLARK: Lana didn't have anything to do with this.
SHERIFF ADAMS: You know, unless Miss Lang has grown several inches and developed an affinity for plaid, this would be a good time for you to pipe down.

'Crisis' demonstrated one of the rules that has governed *Smallville* from its inception — something good can always be made better, and it doesn't matter if it's done at the last moment. The cliffhanging final scene, with Lionel Luthor putting the gun in his mouth and preparing to pull the trigger, was a last minute addition, filmed during the shooting of 'Memoria', a mere week before 'Crisis' was broadcast for the first time.

"That's the best moment of the whole show," executive producer Al Gough maintains. "The rest of the show is a 'box' episode, and one that the fans certainly liked. It has a good contained premise — Lana dies in the teaser, [then you] figure out how she dies — and the bullet-time effect was very cool."

Opposite: Adam on the warpath.

Mark Verheiden recalls that there was considerable discussion in the writers' room

DID YOU KNOW?

After completing work on *Smallville,* Françoise Yip crossed Vancouver to appear as Virago in *Blade: Trinity.*

during the development of the story. "The interesting thing is that there's an inexorable nature to it," he points out. "Whatever Clark did didn't matter, because he finally ends up in the same place trying to help her. If you look back from the end, Clark was wildly ineffectual, because everything he did didn't mean anything — but frankly, that's the fun of that kind of episode. It's more of a sci-fi concept than most that we've done."

"'Crisis' was a difficult one to shoot," Kristin Kreuk recalls. "We did a full day-and-a-bit of running around in this little warehouse, and then continued at night. It wasn't warm yet at night in Vancouver, and I was lying on concrete!"

Mat Beck flew up from Los Angeles to work on set with John Wash and John Hahn to achieve the bullet-time effect. "Once we saw the place where it was being shot, we modified what we'd designed a little bit," he recalls. "Ken Horton wanted to see the bullet really big, going through the rain drops, so when we shot it, we built in a flash zoom so that the camera does a tour around Lana. Lana's running away, you see Adam shoot, and it becomes a frozen moment. We see the bullet travel toward her, through the suspended raindrops, then we see the victim. We come around her, and the bullet's just about to hit her, then we see Clark intercept and step in-between them, and deflect the bullets. Then we go back into real time, and Lana finishes falling over.

"The shot involved a lot of computer-generated material: we had to track the alley and all the dumpsters so that we could have CG raindrops bouncing off them. We had CG raindrops suspended and being intersected and smashed by the CG bullet, so they

Below: Lana desperately calls for help.

SMALLVILLE TORCH

DEAD AGAIN

Teen Dies Twice In Six Months

Adam Knight died last night. Again.

As meteorologists had predicted, the rains came Wednesday — and with them, a deluge of murders. Eight homicides, in fact, courtesy of Knight, 17. This includes the death of Dr. Lia Teng, a world-renowned molecular biologist from Metropolis University. Dr. Teng, whose tenure was yanked for conducting studies without FDA approval, is linked to corporate titan LuthorCorp.

A specialist in hematology, Dr. Teng is rumored to have concocted a serum of minute amounts of cells resembling blood platelets. And we're not talking about playing with lab rats either, folks.

Although no arrests have been made, the FBI is spearheading an investigation into LuthorCorp's association with the serum lab and Dr. Teng's employment. Also up for grabs is Knight's apparent interest in Lana Lang's relationship with Clark Kent.

Lang admits her relief that the saga appears to be over. "I know what it's like to be really scared, and it sounds weird, but that fear put my whole world in perspective for me. I realized what's important in life, and to me, what is important in life is really living."

As for Adam Knight, there's no doubt he would agree that when it comes to living, first time's the charm.

By Chloe Sullivan

had to splash in an interesting way. With Lana, all we had to do was remove the little support that Kristin was leaning on, so we CG'd the bottom part of her body, grafted that onto the real upper part, and created CG hair trailing out behind her, so it looked like she was frozen. Then the instant it becomes real time, she was a complete CG figure who then fell out of frame. So we threw a lot of 'tricks' at the scene, but the upshot was that it ended up looking convincingly like reality. Or maybe 'hyper-reality', since in reality people don't run faster than speeding bullets!"

LANA: Maybe it's time to stop being so afraid. I've always thought that my life is so mapped out for me, but you know, maybe we have more control over the future than we think.

John Glover recalls Miles Millar, Greg Beeman, and James Marshall all being involved with shooting the cliffhanger sequence, and the pick-up at the start of 'Legacy'. "At one point all three of them were back by the monitors," he says. "That's when it gets really exciting on *Smallville*, when we get those creative juices really flowing from the directors and writers and actors all together. Greg came up with the idea of the beginning of 'Legacy', with the phone ringing and my putting the gun down — he said we needed to do this to be complete. It wasn't actually written, but we knew we couldn't leave it like that." ■

LEGACY

WRITTEN BY: Jeph Loeb, Alfred Gough & Miles Millar **DIRECTED BY:** Greg Beeman	**GUEST STARS:** Christopher Reeve (Dr. Virgil Swann), Daniel Bacon (FBI Agent), Gary Hudson (FBI Agent), Timothy Paul Perez (Agent Tobin)

DID YOU KNOW?

Back in 1980, John Glover and Christopher Reeve appeared on stage together in a production of Anton Chekhov's *The Cherry Orchard*.

The key to Clark's spaceship emits a hum, which makes Jonathan behave oddly. The key opens a portal in the cave walls, but before he can investigate, Clark is interrupted by Lionel. Lionel later offers Dr. Swann the conservancy rights to the caves if he'll convince Clark to give him the key. Swann suggests a possible arrangement. The farm is raided by Lionel's tame FBI agents, who are looking for the key, but Jonathan has it with him. Clark overhears Lex talking to his FBI handlers, and Lex admits he's working with them. He also tells Clark that Swann has talked with Lionel.

Clark confronts Swann, who admits his deal and shows Clark a message saying 'I am waiting' — which is meant for Jonathan, who orders everyone out of the caves at gunpoint, except Lionel. The two men fight, and the key starts to shine brightly, returning to its place in the wall. Clark arrives in time to prevent Jonathan from killing Lionel. Lionel lets Jonathan leave, but the wall has sealed up, with the key apparently inside. But it's not there — somehow, it is now in the possession of Dr. Swann.

Lana and Clark grow closer, and she asks Lex for his advice. As a result, she tells Clark that she wants to figure out what she wants from life before getting involved with him again.

LANA: Clark, it's gonna be okay. There isn't anything that you can't recover from.
CLARK: What if you screw up something so badly you can never get it back?

'Legacy' will always be remembered for Christopher Reeve's second and final performance on *Smallville*. "I thought Chris was terrific in this episode," executive producer Al Gough says. "He was in very good spirits, and I think his performance is stronger than in 'Rosetta'. I don't know if the scenes themselves are stronger, but I think his performance is much better — and the scenes with John Glover are really terrific. There's a lot going on there, and it reminds you just how good an actor Chris was. 'Legacy' had pools of emotion, and a lot of really terrific performances: Tom was great, and Schneider did a really good job, too."

Dr. Swann's scenes were filmed in the rotunda at the New York Public Library, which gave them an impressive scope. The situation was unusual, because Reeve's schedule meant that his scenes for 'Legacy' had to be shot about a month ahead of the rest of the episode, before the rest of the script was written, let alone finalized.

Opposite: Clark is displeased by Dr. Swann's explanation.

Tom Welling admits that he felt more comfortable working with Reeve this time around. "Partly it was because it wasn't the first time I had to fly to New York to work

DID YOU KNOW?

At the time of his death, Christopher Reeve was directing the film *Yankee Irving*, a commitment which had prevented him reprising the role of Dr. Swann for the season four première.

and then fly back," he says. "I knew a little bit more about what was in store. It was cool, because when I saw him we were saying, 'Hey, how're you doing?', whereas before there was the whole introduction that you have to have. The first time you meet a prolific person and prolific actor like Christopher Reeve is intimidating. You're in the presence of a true-life hero. The second time, there's that familiarity when you can get along together, so unlike the five or six rehearsals it took for us to get comfortable last time, maybe it was one or two, and we got to raise the stakes a little bit."

"Every word I try to find to describe the experience seems so trite, but it was astounding," John Glover adds. "Chris was just as sharp as a tack — he had a great sense of humor, and was focused and full of energy. He was only interested in doing the best possible job with the part and the scenes that he could have done. He had three scenes: one with me, one with Tom, and one alone, and he'd worked twelve hours the day before. It was a long day for him, but he was tireless — always a step ahead of everybody, because he'd been a director himself, so he had opinions about the best way to do things, which several times were used. He told me that he had felt much better when he was off-camera during his scenes with Tom, but said, 'I'm just here living, and if

Smallville Ledger

LOCAL FARM RAIDED

FBI or Home Invasion?

Last night, Smallville resident Martha Kent was either on the receiving end of a legitimate FBI search or the victim of a rather elaborate home invasion. Either way, a group of people entered the Kent farm property and turned it upside down. Martha Kent was home alone during the event.

Mrs. Kent remembered, "About five dark sedans with black and white government license plates drove up our driveway. A man identifying himself as Agent Tobin shoved a search warrant in my face and then around 15 agents descended on the farmhouse and barn. They destroyed everything and then left. It looks like a tornado went through this place." Visibly shaken, Kent surveyed the room and added, "Just look at our home. It's a mess."

But federal authorities deny any involvement with the Kent search. Sheriff Adams contacted the local bureau chief, who insists agents did not search the farm. Adams said, "Mrs. Kent showed me their 'search warrant.' It looks authentic. It's even supposedly signed by a judge. The problem is, the judge's name doesn't match up with any judge in our county. My preliminary finding is that it's a fake. I hate that someone did this on my turf…"

So, if the FBI didn't search this property, who did and why? "I don't know," stated Sheriff Adams. Then she turned back around and added, "Yet."

By Kathy Romita

Above: Dr. Swann makes his plans.

somebody wants to come and photograph me while I'm doing it, that's fine with me.' Everybody paused, and he went on, 'At least, that's what I try to strive for.'"

JONATHAN: This is your last warning. Stay away from my son.
LIONEL: If you'd raised your son the way I raised mine, maybe you wouldn't have to protect him. You know, weakness isn't something you're born with. You learn it. And Clark learned his from you.

Glover was less comfortable when it came time to shoot the fight between Lionel and Jonathan in the caves. "I was really dreading that," he admits. "I'm a talking actor, so I felt very insecure about coming in to do that scene, with stuntmen and all kinds of physical stuff. But I had such a good time. Schneidy and I really bonded that day, rolling around on top of each other in the dirt!"

"The confrontation between Lionel and Jonathan was something that John Schneider had mentioned in an interview, and I'd always liked the sentiment of it," Al Gough says. "I knew that when we had an episode where we had the two of them together, it would be great to try to use that. 'Legacy' was obviously coming down to that, so we were able to incorporate it into the story." ■

SMALLVILLE MUSIC

'Once In A Blue Moon' by Edie Brickell
'Mona Lisa' by Grant-Lee Phillips
'Love's Divine' by Seal

TRUTH

WRITTEN BY: Drew Greenberg DIRECTED BY: James Marshall	GUEST STARS: Linden Banks (Jonah Doyle), Samantha Banton (Rebecca), Gillian Barber (Mrs. Taylor), Ryan Steele (Doug), Victoria Anderson (Hospital Receptionist), Ivan Cermak (Intern), Mark Houghton (Lab Technician)

DID YOU KNOW?

One of the meanings of the Latin word 'levitas' is actually fickleness – quite a long way from 'truth'.

After sneaking into a LuthorCorp lab, Chloe inhales a gas, which turns out to be a drug that forces everyone to tell her the truth. Pete and Martha almost give away Clark's secrets. Teacher Mrs. Taylor admits she's really a fugitive from justice, and Chloe prints the story in the *Torch*, making herself very unpopular at school. Lex tells Clark that whoever was exposed to the gas is in serious danger. Lana admits to Chloe she is going to school in Paris. Clark tells Lex that Chloe was the burglar, and Lex asks Chloe for her help in bringing down Lionel.

Clark and Pete persuade Chloe to meet them at the hospital, where Pete admits that he's in love with her before she collapses. Lionel visits her, and admits his guilt in his parents' deaths, which Chloe records and sends to her voicemail. Lionel offers to reinstate Chloe's father if Chloe will find out where Clark comes from. Clark finds a potential antidote to the drug as Chloe drives to the farm. Mrs. Taylor's son Bill tries to kill Chloe, but Clark arrives in time to save her and administer the antidote. She admits she was going to ask the Kents about Clark. Lana tells Clark her plans for the future, and Chloe discovers that Lionel has erased the voicemail and now only she knows the truth…

CLARK: Congratulations, Chloe. Within twenty-four hours you've managed to tick off the entire school.
CHLOE: Well, you don't earn your press pass by making friends.

'Truth' was a tour-de-force performance for Allison Mack as Chloe Sullivan. "We see Chloe's dark side," Al Gough notes. "When she can have the truth, will she go for it? It's a very good story for her, and all-in-all a strong episode."

"I was really concerned with 'Truth', because I thought it made her look pretty heartless," Allison admits. "I don't think she is a heartless person, so that was difficult. James Marshall was directing, and he's wonderful, and Tom was very involved with that episode because it was such a risky story for my character. We spoke about it a lot, and actually reshot the last scene because it ended up not being what anybody wanted. It was a really delicate balance because we had to make it incredibly obvious that it wasn't Chloe, it was the drug that was responsible, just so she didn't lose all of her compassion — that was the biggest obstacle to overcome."

Allison deliberately played Chloe differently from normal, to accentuate the difference between her normal and drugged selves. "I consciously tried to play her like a sleazy lawyer," she says. "She was a bit too antsy all the time. I don't know if it read that way or not onscreen, but that's what I wanted."

Opposite: Clark catches Chloe as she succumbs to the drug.

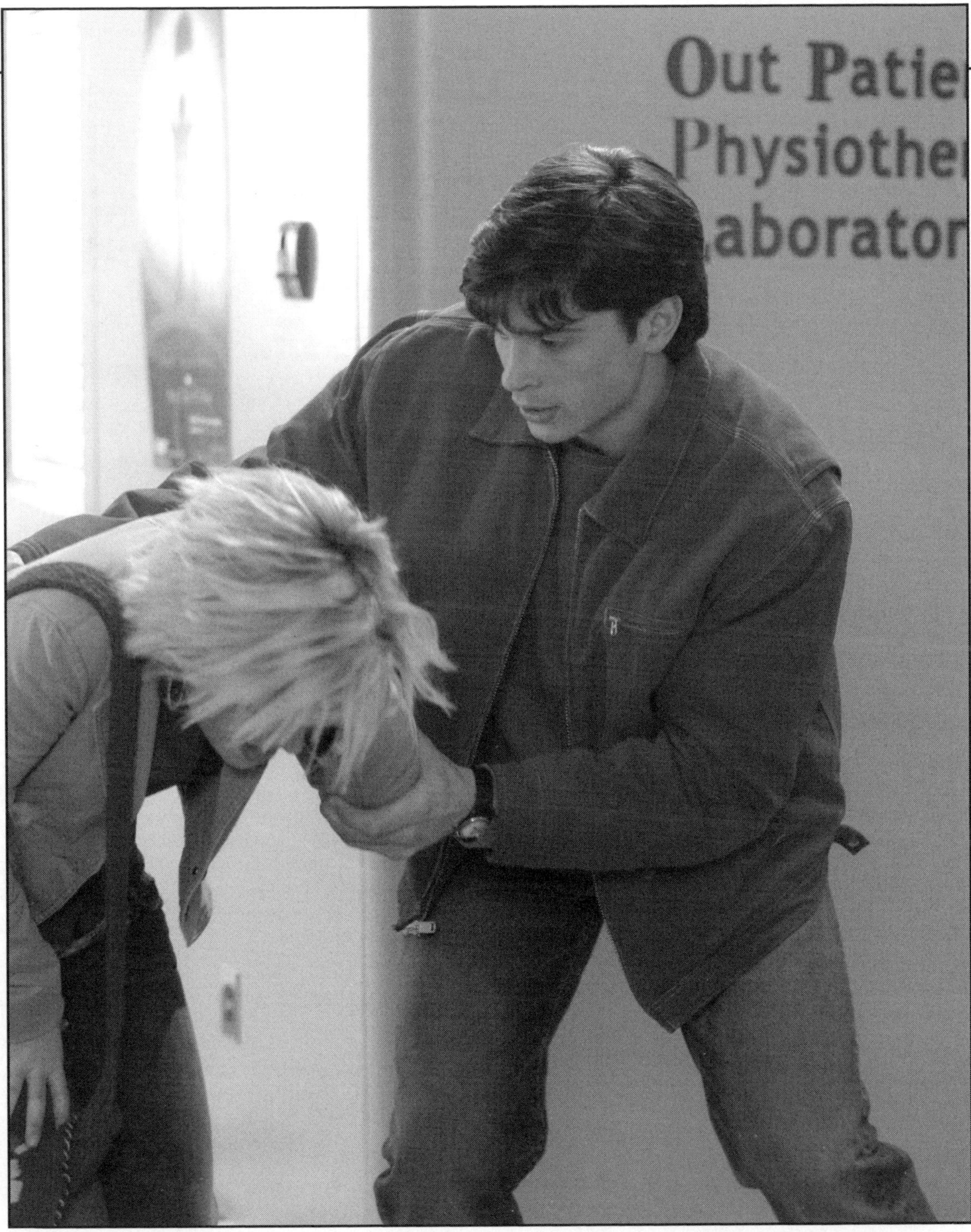
Out Patie
Physiothe
aborator

SMALLVILLE MUSIC

'Someday' by Fastball
'Better Don't Do' by Ing
'Stabat Mater' by Paul Schwartz
'Selling Out' by The Fuzz

She laughs when she considers that, for one week only, she was the Freak of the Week. "It was fun to do something a little bit different," she continues. "We made her like Nicole Kidman's character in *To Die For*, so she became a ruthless, hungry reporter. She was also like an agent — intense, pushy, and aggressive."

CLARK: Why didn't you tell me sooner?
LANA: Because I didn't want anyone to talk me out of it. And the truth is, I knew you were the only person that could.

The first revelation made to Chloe was originally going to be rather different. Instead of the high school teacher being on the run because of something she had done many years earlier, the script centered on a relationship she was having with a pupil. "That was something that was going to require more scenes," Al Gough recalls. "I think that the solution that we came up with was easier. Someone who was hiding from the law was thematically better, and something you could explain in one scene, so you could spend more time with our people."

SMALLVILLE TORCH

THE TRUTH HURTS

While spelunking LuthorCorp labs on an unrelated story Wednesday night, *Torch* editor Chloe Sullivan was exposed to a mystery chemical that forced the truth out of anyone who spoke to her.

So what chemical agent could allow a young woman to gain access to everyone's deepest, darkest secrets? Sergeant Jonah Doyle worked on the Levitas Project, a military operation directed to create a truth serum. "The agent was intended to be the ultimate interrogator," Doyle said. "The truth is a powerful tool. It's a psychological WMD (Weapon of Mass Destruction) and could conceivably be more frightening than nuclear war to some people." But when the toxicity levels proved fatal, the Levitas Project came to a quick end.

Sullivan received an injection of the untested counteractant before the chemical ended her life. The Levitas effects have since disappeared. About her journalistic buzz, Sullivan said, "I just wasn't myself." She refused to comment further.

According to Doyle, LuthorCorp was the civilian contractor of the product five years ago. "We ended that project long ago when it proved ineffective," said LuthorCorp titan Lionel Luthor. But if the project was terminated, Sullivan's exposure proves that the chemicals are still being processed in some capacity at the plant. It also raises suspicions as to what plans Luthor had for the chemical.

By Clark Kent

"The teacher having an affair with a pupil was one of the first thoughts we had," Mark Verheiden recalls, "but we decided it wasn't right for our school. It was frankly a rather unpleasant place to go on a show that's lighthearted, having that at the center of it. The story came out of a retreat we all went on, where we went away for a day and sat around trying to come up with ideas and notions for stories."

Above: Pete reveals his true feelings for Chloe while under the influence.

One of the key character moments, which John Glover recalls as a "fun scene", saw Lionel still able to force Chloe to do what he wanted, even though, like everyone else, he had to tell her the truth. "Lionel is such a bastard that he could manipulate her even then," Verheiden says. "He told her a selective truth!"

'Truth' marked Verheiden's final input to the series, after being involved with *Smallville* since early in the first season. "I just love Chloe," he says. "Lana's the unattainable girl, but Chloe's who you probably think or hope you are: a beautiful friend who's witty and quick, and always there at a pinch. She's a great character, and it was great to go out on a story about her." ■

DID YOU KNOW?

Gillian Barber (Mrs. Taylor) is also known for her role as Ambassador, later First Minister, Dreylock on *Stargate SG-1*.

MEMORIA

WRITTEN BY: Al Gough & Miles Millar DIRECTED BY: Miles Millar	GUEST STARS: Paul Campbell (FBI Agent), Michael-Ann Connor (Receptionist), Martin Cummins (Dr. Garner), Wayne Dalglish (Young Lex Luthor), Alisen Down (Lillian Luthor), Matthew Harrison (Tech), Terence Stamp (Voice of Jor-El)

Lana finds Lex, in the grip of a nightmare, standing on the ledge of a balcony at the Luthor mansion. Lex was dreaming of his youth at a private school where his father found him cradling a blanket he called Julian, his dead brother's name. It is revealed that Lex is undergoing treatment at Summerholt, where Dr. Garner is regressing him in an effort to recover the memories he lost through the shock treatment. Clark asks Lex not to proceed, and even asks Lionel for help. Lex remembers an argument between his parents regarding him and Julian, and recalls Lionel finding him standing over his brother's crib, saying, "It was an accident." When Lionel saw that Julian was dead, he slapped Lex.

Smallville Ledger

SUMMERHOLT FOUNDER COMATOSE AFTER MEDICAL ACCIDENT

Known for its sleek, modern appearance and cutting-edge medical practices, the Summerholt Neurological Institute is now the center of a police investigation. After several small explosions, Chief Neurologist Dr. Lawrence Garner, 34, was discovered unconscious beneath fallen wires, broken ceiling tiles, and flying sparks. Metropolis police found a shattered water tank in the center of the facility and green water everywhere.

Investigators report that the alarm in the "memory well" triggered the series of explosions. But Summerholt employees refuse to comment on what the building's floor plan denotes as the "memory tank". Dr. Garner, who is still comatose at Metropolis General, has a reputation for pushing the boundaries of science.

Dr. Garner's phone records connect him to an elite patient roster. However, an incoming call from Lionel Luthor, the only non-client, was logged 30 minutes before the facility meltdown. Witnesses saw an argument between Luthor and son Lex in the lobby just moments before the first explosions.

Building manager Darlene Simon claims, "Summerholt was founded by Dr. Lawrence Garner and has no affiliation with LuthorCorp or the Luthor family" — curious, since Lex Luthor's name was found scribbled in the building's sign-in registry throughout the last two weeks.

By Sandy Maggin

SMALLVILLE MUSIC

'My Immortal' by Evanescence

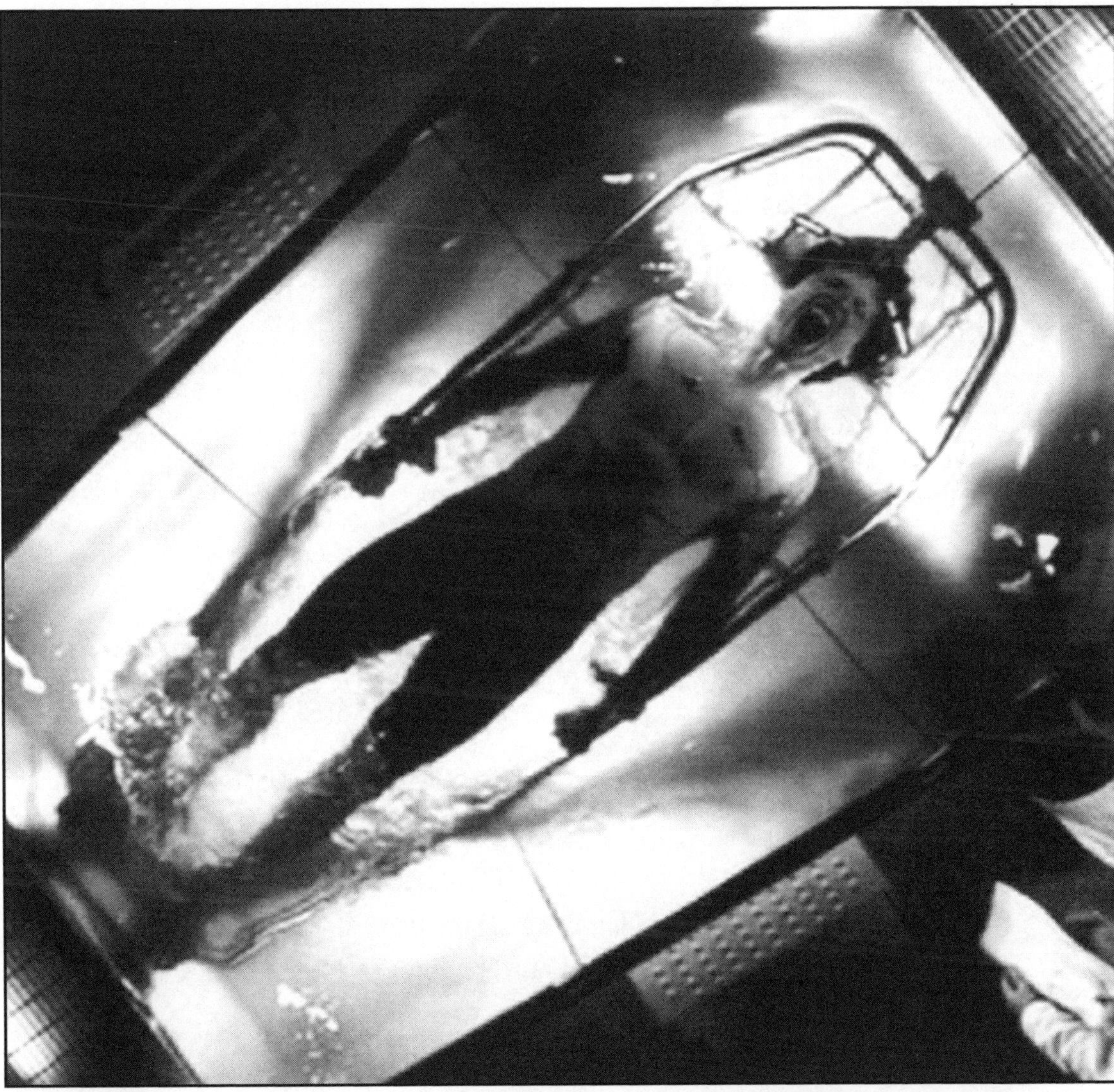

Lionel suggests that Clark should try to persuade Dr. Garner not to continue. However, the kryptonite in Garner's experimental tank affects Clark, and Garner wires him up for regression. Clark remembers his parents Jor-El and Lara placing him in the Kryptonian spacecraft. Clark's screams feed back into the machinery, making it explode, and Lex comes to Clark's rescue. Clark asks Lex why his father hates Lex so much, which triggers a memory for Lex of discovering his mother just after she had smothered Julian to death. Lex tells his father the truth — even as a child, he knew that Lionel would never hurt his heir, so he took the blame. Later that night, Martha tells Clark that his first word was "Lara". ■

(See page 102 for an in-depth look at the making of this episode.)

Above: Clark relives his memories in the Summerholt memory well.

TALISMAN

WRITTEN BY: Ken Biller
DIRECTED BY: John Schneider

GUEST STARS: Nathaniel Arcand (Jeremiah Holdsclaw), Gordon Tootoosis (Prof. Joseph Willowbrook), Teryl Rothery (Real Estate Agent)

DID YOU KNOW?

This isn't the only time Lionel is seen in a stretch limo: in the concluding chapter of *The Chloe Chronicles II*, he meets Chloe and Pete in one.

Lionel's team find an artifact in the caves, but Lionel won't let Professor Willowbrook and his student, Jeremiah Holdsclaw, study it. The artifact breaks open to reveal a blade that throws energy into Jeremiah, allowing him to resist the impact of bullets and imbuing him with superspeed. Willowbrook thinks the artifact is the Starblade, which gives its bearer powers to protect the blade until it can be delivered to Naman to use against his enemy, the Zegit. Lex offers to help Lionel track Jeremiah, and despite his father's refusal, Lex still follows Jeremiah's trail. Willowbrook tells Lionel that the blade will crumble if the Zegit touches it. Jeremiah believes he is Naman, and Lionel is the Zegit. Jonathan suggests using kryptonite on Jeremiah, but it doesn't work, and Jeremiah injures Clark with the Starblade. Only an inexplicable energy from Jonathan's hands seals the wounds. Clark battles Jeremiah just as he's about to kill Lionel, leaving Jeremiah in a coma and Lex able to free his father. Clark can't see which Luthor causes the blade to crumble by touching it.

Lana is accepted into the Paris School of Arts. Lex puts the Talon up for sale to release money for her. Pete admits to Chloe that he hasn't been home since he learned that his parents are divorcing, and that his mother is possibly taking a Federal judgeship.

CLARK: Professor Willowbrook thinks that I'm Naman. Jor-El wants me to be some sort of Kryptonian warrior. I just want to be Clark Kent.

For the second episode in succession, the cast were directed by someone with whom they were very familiar, on this occasion Jonathan Kent himself, John Schneider.

Schneider's directing experience stretches back to his time as one of the *Dukes of Hazzard*, and with his in-depth knowledge of the various sets and the characters after three years working on the show, he was an ideal choice for director. "John Schneider did a good job with a not-very-good script," executive producer Al Gough says. "It's one of those episodes where you'd like to go back and redo it, so it's not an episode about two older men, Willowbrook and Lionel, trailing around looking for a knife. If you'd never watched the show before, you'd have no idea what they were talking about. There was no entry point into that episode. We got too mythologized, and there was nothing for a younger audience."

Opposite: Professor Joseph Willowbrook.

The cast found it interesting to have Schneider behind the camera. "John has a great familiarity with the crew and actors," Tom Welling admits. "For me, a lot of the time

Above: Lex and Lionel examine the blade.

being on set is about environment, and if you can create a stable, comfortable environment for everybody to work in, then you can have great things happen — I think both John and Miles have that."

"John Schneider had a really great time directing, from the look of it," Kristin Kreuk comments. "He'd wanted to do it for such a long time, and it's great that he got the opportunity. A lot of that episode got cut, though — there was a big party sequence in there for Clark's birthday." (Happily, it didn't stay on the cutting room floor forever: the scene is included as an extra on the season three DVD box set, along with an extra confrontation between Holdsclaw and Lionel.)

LEX: Who are you? Are you a small-town girl content to spend her days gossiping in a coffee shop with her friends? Or are you the girl who goes to Paris?

DID YOU KNOW?

Shortly after working on 'Talisman', John Schneider appeared in the earthquake-based mini series *10.5*.

John Glover found it interesting being directed by his fellow actor. "Because our characters are such rivals," he notes, "there were certain choices he made as director that I could see he was making more as Jonathan Kent than John Schneider. I'll explain what I mean. There was a scene where I went to the university to talk with the professor, and they had a car waiting for me. They rented a huge stretch limousine. I kept thinking that Lionel wouldn't ride in them: I see Lionel as being powerful, but without having to be flashy. But John was dwelling on Lionel's power. It was little differences like that which made it interesting, because I could see he was seeing Lionel from Jonathan's point of view — and I'm sure I would do the reverse if I was directing an episode."

Schneider was pleased that he was given plenty of time to prepare his episode as he followed the development of the script. "I saw all the different versions of what that story was," he explains. "We got to do some interesting things, some of which met with a bit of reaction on some people's parts, but I really liked it. I believe that blocking can actually help your scene. If you're doing a scene about separation, then physically separate the people onscreen; if you're doing a scene about intimacy, put them in an elevator. We did some separation scenes between Lex and Lionel that I think were quite interestingly blocked."

Al Gough was particularly pleased with the final scene of the episode. "I think people will talk about that scene, because it's the perfect super-villain justification," he says. "You think you're the good guy. Lex makes a good point. If an alien came to Earth, whoever had to deal with that person would be a hero. It's how he justifies things to himself." ■

SMALLVILLE MUSIC

'Chase Me' by Katie Herzig
'Reason Why' by Rachael Yamagata
'From Afar' by Red Letter Day

Smallville Ledger

PROFESSOR LAMENTS LUTHORCORP ACTIVITY

Radical Kawatche Student Promises Action

An ancient Kawatche legend tells of a giant with a heart of stone encased in ice who came from far away and usurped the natives' land for many seasons. No attacker, no matter how strong or cunning, could defeat the greedy invader. After the people had suffered for longer than anyone could remember, one warrior prayed to the Great Spirit to warm the giant's heart. As the Spirit complied, the giant's blood melted and flowed out, forming the chilly Elbow River, and his internal organs were crushed by the weight of his corrupt heart. Thus did the land's rightful owners prevail through patience, not violence.

Today, Professor Joseph Willowbrook likes to tell that tale to illustrate the present-day plight of the Kawatche.

"Mr. Luthor knows that what he's doing is wrong," expounds Willowbrook. "The caves have belonged to my people for countless generations, and now we must beg permission to even go inside them. But I suspect that as in the legend, the time will come when all will be set right again. Whether Mr. Luthor can survive the consequences of his cold-heartedness remains to be seen, but we will wait for that day."

However, some of Willowbrook's followers refuse to adopt his Gandhi-like determination, preferring action to passive resistance. Jeremiah Holdsclaw, a teaching assistant of Willowbrook's and a full-blooded Kawatche, vows to retake what he considers his birthright: "Yes, we are a peaceful, law-abiding people, but sometimes change can only be effected with force. Patience is something only our forbears could afford — our heritage is being ransacked while we stand idly by."

By Christopher James Beppo

FORSAKEN

WRITTEN BY: Kelly Souders & Brian Peterson
DIRECTED BY: Terrence O'Hara

GUEST STARS: Neil Flynn (Pete Dinsmore), Amber Rothwell (Emily Dinsmore), Gary Hudson (FBI Agent Frank Loder)

DID YOU KNOW?

Rather ironically, given he's playing a scientist who created a clone of his daughter, Neil Flynn (Pete Dinsmore) had voiced characters for the 2002 MTV animated series *Clone High*, about a group of clones of the world's most powerful people.

Lex gives Lana an open-ended ticket to Paris, and advises Clark to be honest with Lana if he wants her to stay. Emily Dinsmore, the clone of Lana's childhood friend (from 'Accelerate'), has now grown into an adult, and escapes from her prison. She kills the new owner of the Talon and takes his place. Emily, claiming to be the new owner's daughter, asks Lana for help with the Talon, and then knocks her out. When Lana doesn't turn up for dinner, Emily tells Clark that Lana has already left for Paris.

When Lex tells him that the new owner doesn't have a daughter, Clark confronts Emily, but she escapes. Lionel tells Chloe that Emily has escaped, and Chloe and Clark deduce where she's hiding. The two then rescue Lana.

Lionel tells Lex he knows about Lex's deal with the FBI; however, his own contact, Loder, confirms they have no evidence against Lionel. Chloe tells Lex about the erased voicemail that is still somewhere in cyberspace. Loder beats up Pete, trying to find out more about Clark, but Pete refuses to talk. Lex gives Loder the transcript of Chloe's voicemail to get him back working against Lionel. Pete decides that he can't cope with knowing Clark's secret and leaves Smallville with his mother. Clark decides not to tell Lana his secret. The FBI arrest Lionel for the murder of his parents.

CLARK: You sound like Lex. He thinks I'm the reason Lana's leaving.
PETE: Well, I hate to agree with Uncle Fester, but I don't think Lana's going to Paris just to see the Mona Lisa.

Like many of *Smallville's* best episodes, 'Forsaken' takes one of the potential situations facing the lead characters and dramatizes it in an extreme form. "Clark would like to keep Lana in a box, and have nothing happen to her," Al Gough explains, "but can't do that. The Clark/Lana story is very good in this episode, and we saw some Clark/Lex friction over Lana's departure."

Inevitably, the penultimate episode of the season sets up the various plot points for the finale, particularly the feeling of isolation that Clark has in 'Covenant', when everyone appears to have betrayed or deserted him. The first departure Clark was aware of was Lana's trip to Paris. "During the third year, Lana spent the first third trying to understand Clark, and waiting for him to come around," Gough says. "Then the second third she was in this not very good relationship with Adam, and then the final part was her deciding that 'I'm not going to define myself with a guy, I want to see what else is out there.' She was trying to expand her horizons — the small-town girl leaving the small town, thinking she's leaving these problems behind her. And of course, when she

Opposite: Emily Dinsmore is *not* happy.

SMALLVILLE MUSIC

'When The Sun Goes Down' by Charlie Mars
'What You're Thinking' by Christopher Jak

gets there, they come back in an even better way! I think Lana's arc in the third year was very good. The Clark/Lana relationship elements become a little stale, but she's the only normal character on the show, a normal teenage girl. Clark does all these things to her, and the audience understands why, because they're in on the secret and she's not. She reacts perfectly normally to a guy who's one minute the nice guy, and the next is 'motorcycle-riding jerk'; all these things happen that she can never understand, and he never tells her."

LIONEL: I did it to make you stronger.
LEX: The hell you did! You can't keep hiding every deceitful thing you've done to me under the guise of mentorship! You brought this on yourself.

The second, and more unexpected departure, comes when Pete decides that he can no longer cope with the pressures of knowing Clark's secret. His leaving was set up in the scenes between him and Chloe when they discussed his parents' divorce, broadcast in 'Talisman' but actually filmed as part of 'Legacy'. "Due to time, that entire subplot got lifted out," Gough explains. "Quite frankly, it fit better in 'Talisman' because it was a better wind up. If we had set it up in 'Legacy', by the time you got to 'Forsaken', you wouldn't remember it."

Below: Lex watches coldly as the FBI arrest Lionel.

Smallville Ledger

LIONEL LUTHOR ARRESTED FOR MURDER

In a completely unexpected turn of events, FBI special agents arrested LuthorCorp CEO Lionel Luthor this week, charging him with first-degree murder. Even more shocking was the identity of his alleged victims: Lachlan and Eliza Luthor — his own parents.

The elder Luthors perished decades ago in a tragic tenement fire in the 'Suicide Slum' area of Metropolis. While arson had been suspected, previous police inquiries turned up no solid clues pointing to foul play.

Now, however, federal prosecutors believe they have enough evidence to bring a case against the famed titan of industry, although they declined to provide reporters with details about what, if anything, links Lionel to the heinous crime. "We have no plans to try this in the media," said a government spokesman. "We'll save it for the jury."

Sources within the FBI's Kansas field office confirmed that the crux of the investigation rests on claims that young Lionel hired would-be crime boss Morgan Edge to actually carry out the murders, supposedly so Lionel could inherit his parents' life insurance money.

A notorious figure in the Metropolis criminal underworld, Edge was said to have died last year following a mysterious, still unexplained explosion in Smallville. "It's unfortunate that law enforcement officials were never able to make charges stick against Edge while he was alive," Taylor said, "but they shouldn't take their frustrations out on an innocent man like Lionel Luthor."

By Frank Moore

Gough acknowledges that with everything else going on in the episode, Pete's leavetaking of Clark wasn't perhaps weighted with as much importance as it should have been. "It wasn't meant to feel like a footnote, but it does a bit," he says. "Pete deserved a better send-off. We think that Sam is a terrific actor, but we felt that the character had played its course. We decided it was time. We didn't want to kill Pete, obviously, so he's out there and available to guest star or be available for a plotline."

With the tension rising between the various characters, one of the worries that executive producer Greg Beeman had about the kinds of scenes that regularly turn up on the show was alleviated. "How many Clark/Lana scenes in the loft have we done?" he asks rhetorically. "How many Clark/Chloe scenes in the *Torch* offices have we done where they're trying to figure out the mystery? We've done hundreds. One of the tasks and disciplines any actor in a TV series has to face is that you're going to play the same person for years. There's a moment when you discover the character, and then there's the discipline of finding newness and freshness in it. Part of my job is to ensure that the actors don't fall into a habit or give a habitual 'performance'. Whenever they get a chance to do something different — whether it's playing a different character like in 'Relic', or showing a different side to their usual character, then it's fun for them to find something new." ■

DID YOU KNOW?

Gary Hudson (Agent Loder) worked with John Schneider on the 1997 *Dukes of Hazzard Reunion!* TV movie.

COVENANT

WRITTEN BY: Todd Slavkin & Darren Swimmer
DIRECTED BY: Greg Beeman

GUEST STARS: Adrianne Palicki (Kara), Terence Stamp (Voice of Jor-El), Robert Wisden (Gabe Sullivan), Gary Hudson (FBI Agent Frank Loder), Brandon McLaren (Delivery Man)

DID YOU KNOW?

In DC Comics continuity, Kara is Superman's cousin, the daughter of Kryptonian scientist Zor-El and his wife Alura. On Earth she adopted the name Linda Lee.

A girl called Kara arrives at the Kent farm claiming to be from Krypton, Clark's home planet, and says she's come to take him back. Jor-El's spirit lives on within the cave walls, and is now ready to tell his son everything. Jonathan admits that his deal with Jor-El was that he would return Clark to his birth father. Lionel sends Clark a key to Lex's secret room, and Lex walks in on Clark just as the latter discovers Lex's investigations into him. Clark forgets his promise to take Lana to the airport and arrives in time to see her give Lex a farewell kiss. Clark gives his evidence against Lionel in court, but tells Lex their friendship is over. Feeling betrayed by everyone, Clark is about to enter the portal in the cave wall when Jonathan tells him that Kara is in fact human, and was taken through the portal during the meteor shower. Jor-El tells Clark to enter the portal, and disposes of Kara. When Jor-El puts an energy rope around Jonathan's neck, Clark agrees. As Clark enters the portal, Lionel instructs the prison barber to shave him bald; Chloe and her father arrive at an FBI safe house, which then explodes; and Lex takes a drink of poisoned brandy.

In the field outside the farm, Martha watches as a Kryptonian symbol is burned into the corn, while inside the cave, Clark crouches naked in a fetal position as Jor-El proclaims that his son will now be reborn…

CLARK: I can't fly.
KARA: Not yet. This is just the beginning, Kal-El. You have no idea how powerful you'll become.

"I think the season finale is terrific," Al Gough says. "The Kara story works really well, the Lex/Lionel story works well, and obviously it's very emotional."

In a story that was filled with stunning images — from the Kryptonian symbol emblazoned in the cornfield, to the look on Lex's face when he finds Clark inside his secret room — one stands out: Lionel Luthor sitting in the prison chair, his locks being shorn. "That was a fantastic last five minutes," Gough says. "That was John Glover's idea. He was doing a play in Philadelphia for the summer and needed to cut his hair and said, 'What if we did it in prison on the show?' We were looking for that device to cut back to, while all these other things were happening. That look he gives at the end is just priceless. He's bald and looks like Lex Luthor in the comic books. It was a great send-off for the season, and allowed us to come back in season four and really shake it up."

"A lot of it came from the fact that I've never played a part for so long," John Glover explains. "I'm used to doing a lot of theater, and even within theater, a lot of repertory, so I'd go from one character one night to another character the next night. When they told

Page 97 (following color section): Clark asks Lionel about the key.

posite: Clark stops a wounded Morgan Edge from escaping plain sight of Lex in 'Shattered'.

Above: A selection of logos and signs produced by the art department for the show.

Next page: Clark's Kryptonian father, Jor-El, takes Lana's great aunt Louise for a romantic moonlit flight in 'Relic'.

COVENANT

Above: Jonathan, Kara and Clark confront Jor-El.

me that I was going to prison, I just thought it could bring out a different aspect of Lionel. We were going into season four, so I thought it could liven things up a bit."

Glover actually made his proposal during the filming of 'Memoria'. "I was in the trailer putting the wig on, and Miles came in to talk about one of the scenes we were doing that day," he recalls. "I asked if he'd be interested in my shaving my head for the season finale. I thought he was going to go through the ceiling of the makeup trailer, he was so happy! That's why this job is so wonderful, because ideas come from all around."

CLARK: Ever since I've met you, I've been defending you, making excuses for you to people like Pete, like my parents. Telling them, "You can trust Lex Luthor. He's a good guy. He's nothing like his father." I was wrong.

"I think Lionel is to *Smallville* what Lex is to the later Superman stories, and the stronger you can establish that, the more drama you can create," Tom Welling reasons. "I like these 'grown-up' storylines. We can sometimes get stuck in the high school mentality. The more grown-up situations we can get into, the more interesting it is, because then, at least with my character, he's a fish out of water. He doesn't yet have the maturity to handle those situations, and I think it added drama to the end of the year."

SMALLVILLE MUSIC

'One Moment More' by Mindy Smith

A lot of elements came together to ensure that the season finale was memorable, including the music. "The last episode of the third season was a very good show," states music supervisor Chris McGeary. "It was a good representation of what Mark Snow does with the score, what I do to bring it together, and what Madonna does sourcing the songs as well. It all worked very well together."

"Sometimes it's amazing the impact you can have with just one song," Madonna Wade-Reed adds. "I'm a big proponent of not diluting the strength of music by having too much. If you have too much, it all tends to blend together."

As the season came to a close, the producers took the opportunity to review the strengths and weaknesses of the previous twenty-two episodes. "The Adam subplot we felt didn't work well," Al Gough acknowledges. "The science fiction elements of the show work best when they're true to Superman or the concept of the show: the meteor freaks who are working out some inner demon, with some emotional component. It got very mechanical, to the detriment of the show.

"But a lot of episodes were some of our best ever. I think that the series completely tracked emotionally from episode to episode. Performances were much better — Tom came in strong, and was great through the whole thing, and it was probably Michael's best season to date. It was a strong season for the Luthors overall." ■

DID YOU KNOW?

At her Comic-Con appearance in the weeks following the broadcast of 'Covenant', Allison Mack teased fans by insisting that Chloe died in the explosion at the safe house.

Smallville Ledger

MASSIVE EXPLOSION WIPES OUT ISOLATED FARMHOUSE

Owners/Occupants Unidentified, Media Access Blocked

A huge fireball sent flames and smoke high into the air when a midsized home on the outskirts of town exploded yesterday. There were no official reports of any injury or death.

While it is common practice to withhold victims' names until bodies are clearly identified and next of kin notified, the authorities are being especially mum on this unexplained occurrence. There was no statement, and news teams were blockaded several hundred feet from the blast site.

Strangely, even hours after the incident, vehicles and personnel from local law enforcement, fire, and emergency departments were still not allowed on the premises. Sheriff Adams was seen angrily confronting three black-suited men near the edge of the press camp, but even she was turned away. Before the airspace was closed, a Metropolis *Daily Planet* news chopper did manage to capture images of the explosion's aftermath, reportedly showing dozens of unmarked black sedans, SUVs, and vans surrounding the smoldering property. After their initial broadcast, those images were embargoed, and a complete media crackdown went into effect.

A preliminary check of deed records did not reveal a property owner for the plot where the house was located. Indeed, all paperwork relating to the structure and surrounding land seems to be nonexistent in the county archives.

Regardless of whether or not the cause of the mysterious blast is ever determined, it seems likely that if anyone were inside at the time, it would be a miracle if they survived.

By Jim "Slim" Bradlee

MEMORIA: IN DEPTH

"We love 'Memoria'. Miles did a fantastic job directing it. Across the board, it's one of the best of the series." — Alfred Gough

For his directorial début, executive producer Miles Millar deliberately chose a story that touched on the mythology of the series, shining light on aspects of major characters' pasts that had not previously been seen. Although earlier episodes had established the fact that Lex had a baby brother, Julian, who had died of crib death (also known as Sudden Infant Death Syndrome), this was the first time that *Smallville* had ventured into that period of Lex's life.

"The first time that Julian cropped up was in the first season story 'Stray'," Gough explains. "We always knew what the story was, and it was just a question of finding the right time to do it. That was something we had up our sleeves. We always wanted people to think that Lex was somehow responsible, but in 'Memoria' you find out that he was covering for his mother, who was a little cracked."

In accordance with their usual policy of only informing the actors of information that they directly needed, Michael Rosenbaum hadn't been privy to the knowledge about the reality of Julian's death when he filmed the scenes in 'Shattered' where he cradled a doll, believing it was Julian. "They didn't tell me until 'Memoria' came up," he recalls.

Below: Young Lex discovers his brother's body.

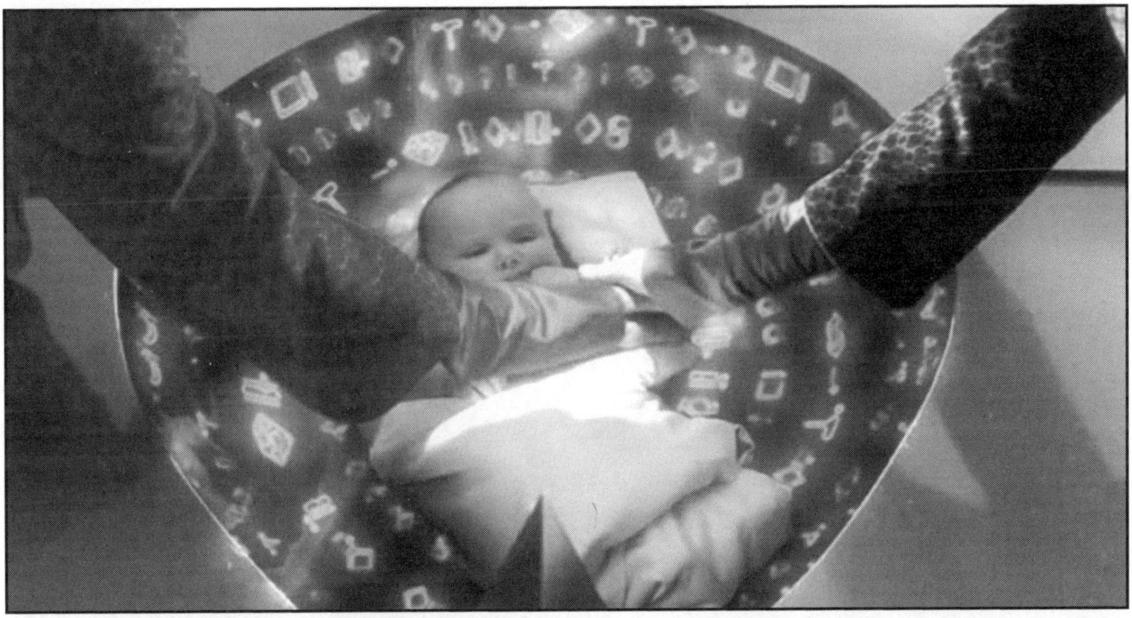

Above: Jor-El and Lara place baby Kal-El in the spaceship.

"I don't know if I even asked. They could have easily given me hints, but they didn't."

"It was something that we always wanted to do," Al Gough continues. "We wanted to see Lex's childhood, and get a glimpse into that. We also saw Clark's earliest memory, of his parents putting him into the ship."

"If I was going to direct, I wanted to do an episode that meant something to the show," Miles Millar explains. "The biggest thing obviously was that it was a good script. The story worked. The challenge for any director on the show is that they rotate, and they're never there for the development. They're given a script and have seven days to prep it and then ten to shoot it. Whereas, as a writer and director, you have the luxury of being able to write your own material, and that's a big help, because you can think more about the scene in terms of blocking and the visuals earlier in the process."

Millar went up to Vancouver for his week of preparation, looking at the sets and working out exactly what he wanted to do. "That was the part I found really challenging," he says. "Actually sitting on the set and blocking out where the actors would go, and the camera moves. Then it became more of a game — having a limited number of shots to get something in the time that you have. You have overhead schematics of the sets and you draw little cameras and stick figures. It's fun. It felt like being back in film school."

The episode originally opened with Lionel arriving at young Lex's prep school to be told by the headmaster about Lex's problems. "It wasn't working," Millar admits. "It was supposed to be a long Steadicam walk-and-talk, and we had huge rain towers up, with tons of extras. We had an hour to get the shot, and we tried many times. It was a second

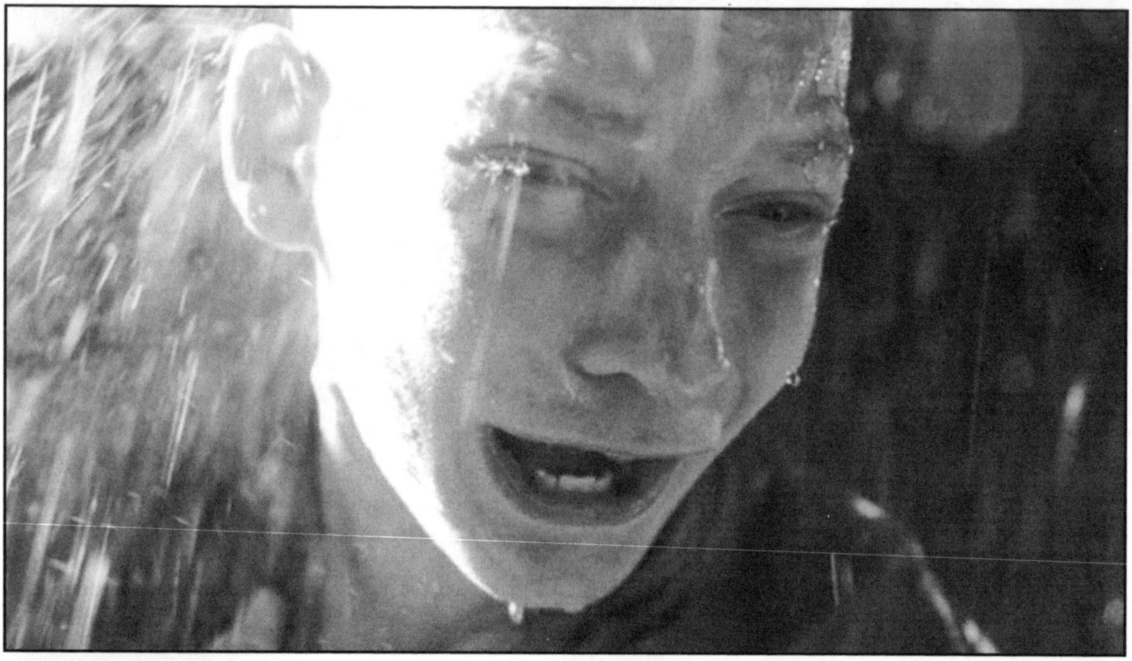

unit day and the Steadicam operator wore glasses, and they got all steamed up in the rain so he couldn't see anything." The scene, filmed at St George's Preparatory School in Vancouver, led into the opening as filmed, where Lionel comes out to find young Lex on the balcony.

LEX: I really don't have to explain myself to you, Clark.
CLARK: Yes, you do, Lex. Because I can't understand why my best friend keeps lying to me.

The younger Luthor was played by Canadian actor Wayne Dalglish, who is also a World Champion in martial arts and weapons — the youngest ever in some fields. He was willing to shave his head to play Lex. "He hadn't done much before," Millar comments. "He's a kid who's home-schooled, completely natural, and unbelievably polite. He was completely professional."

For his scenes as the younger Lionel Luthor, John Glover was delighted with the alterations to his makeup. Rather than the hair extensions he sports normally, "I got a real wig this time," he says excitedly, and he was also supplied with some 'eye pulls', pieces of clear tape used to give a temporary 'facelift'.

Opposite: Lex's flashback transition.

Below: Lionel hands Lex his twelfth birthday present from his mother.

Because the episode was full of transitions between the past and the present, Millar was keen to ensure that they were very clear to the audience. For previous episodes incorporating flashbacks (such as season two's 'Lineage'), various techniques had been used, and in 'Memoria', Millar moved between the two timeframes by elegantly transitioning between the older and the younger Lex Luthor [see opposite]. "Those shots were tough," Michael Rosenbaum recalls. "I had to rely on Miles to guide me through, because I didn't know emotionally where I was meant to be."

The whole shooting schedule of the episode was a learning curve for the novice director and the cast. "It was so good for all of us," John Glover says. "It was good for Miles because he was up on set for ten days to see how it all worked. On the first day, we had a disagreement — he wanted to go one way, and I wanted to go another. In the end, I went his way. I might have been a bit upset about going his way at first, but I went there and then as soon as I did, I realized that he was right. We had a good laugh about it, and I just trusted him completely after that. It was actually good that it happened."

"I was surprised in a good way," Michael Rosenbaum adds. "He didn't come into the episode with an attitude of

MEMORIA: IN DEPTH

Above: Lex and Lionel discuss Lex's unwise course of action.

'This is what I want to do.' He didn't come in as a producer or writer; he came in as a first-time director who really wanted to learn. He knew what he wanted and he had to trust other people. He allowed himself to trust the director of photography and the actors. He learned a lot, and I learned a lot. He wasn't afraid to look like a first-time director, and by the end he didn't appear to be a first-time director."

"They knew I was a first-time director, so there was a certain nervousness," Millar admits. "For me it's all about the fact that they've been playing these roles for three years, and I've also been writing these characters for three years, so when they talk about how well they know the characters, I can say I know them better. The first day, which was the scene with Tom and John Glover, definitely challenged me, but we began to work together, and discussed and changed the blocking. After I gained their trust, it was plain sailing from there on in. They trusted me after day one: they knew I wanted to hear what they had to say, and listen to their thoughts and feelings."

"A lot of small talk got cut out," Tom Welling explains. "With Miles, his knowledge of character, story arc, and where this is all going means all the ideas can be expressed in less words, and be more efficient in a sense."

Millar had a very clear sense of what he wanted the show to look like. "I like extreme lighting," he says. "I like silhouettes and shafts of light. I was particularly happy with the scene in the Luthor library, with Lex and Lionel shot at a really low angle with great colorful shafts of light coming in through the window."

Much of the episode was filmed in an old train-car washing building, in which the tank full of kryptonite-laced water was placed, along with the rest of the paraphernalia of the Summerholt Institute. The exterior and foyer shots were filmed at an unused building belonging to the Sierra Wireless Company, but the main washing building was somewhere that had been spotted on a location reconnaissance earlier in the season, but hadn't fitted with previous plans. The tank itself held four hundred gallons of water. It was designed by production designer David Willson, and constructed by the crew under Rob Maier, in collaboration with Mike Walls, who was preparing the practical effects needed for its eventual destruction. All of this had to be done in the seven-day preparation period for the episode — although an extra day's shooting time was found by producer Bob Hargrove, preparation had to fit within the usual *Smallville* schedules.

The attention to detail applied to all elements of the production. For Lex's twelfth birthday party, the set was packed with mountains of food and gifts waiting for the children who don't turn up, leaving Lex all alone. Millar took his inspiration from the 1942 movie *Kings Row,* which co-starred Ronald Reagan, in which a child has a birthday party and no one comes. Glover enjoyed working on the scene. "It was great to have the chance to play with that wonderful Wayne Dalglish, and go back to the root of what happened, and see where the sickness between Lex and Lionel began," he says. "I could put all that love there in the beginning. There was such love in that

Below: Lex's lonely twelfth birthday party.

MEMORIA: IN DEPTH

scene at the birthday party, where I tell him about the box." For Glover, it was another layer to add to his character. "To be able to play that love from father to son, to have that instead of the difficulties that Lionel has with the grown-up Lex was so great, and I appreciated the chance." The box itself was first seen in season one's 'Metamorphosis': Lex gave it to Clark as a thank-you keepsake, and he used it to keep Lana's necklace out of harm's way.

Another connection to the past surfaced during the confrontation between Lionel and Clark, as the corporate titan shows Clark footage secretly filmed at the Belle Reve asylum. 'Memoria' was an example of the increasing serialization of the show. "It's the nature of television to have callbacks to earlier episodes," Millar explains.

Alisen Down, who played Lionel's wife Lillian in the flashback sequences, was another Vancouver actress who captivated the production team when she auditioned for the role, working in scenes with John Glover. "As small as that role was, she was so stunning with it," Glover comments. Down had the capacity to get into the highly overwrought emotional state needed for the scenes and remain there for the three hours or more necessary to shoot the harrowing arguments between Lionel and Lillian, and then the immediate aftermath of her killing of Julian. The nursery itself was built for the production underneath the balcony at St George's Preparatory School, and dressed, complete with specially designed bunny wallpaper, by Rob Maier.

GARNER: What's the matter, Mr. Kent? You're not looking well.
CLARK: You have to stop what you're doing.
GARNER: You've always been intrigued with my research. Now, you're going to have the opportunity to be a part of it.

For the very end of the scene, where Lex discovers the dead body of his brother, a prosthetic doll was prepared. There were considerable discussions as to whether it could be shown onscreen, and executive producer Ken Horton went to bat with the various executives to ensure that it would. In case he was unsuccessful, Millar had filmed various smaller shots, close-up on the baby. "I always got coverage," the director explains. "With coverage you can protect yourself, because you never know what you're going to need. As producers, one of our primary jobs on the show is to try and fix in

Opposite: Clark in the tank (above), and the aftermath of Lex's destruction of the tank (below).

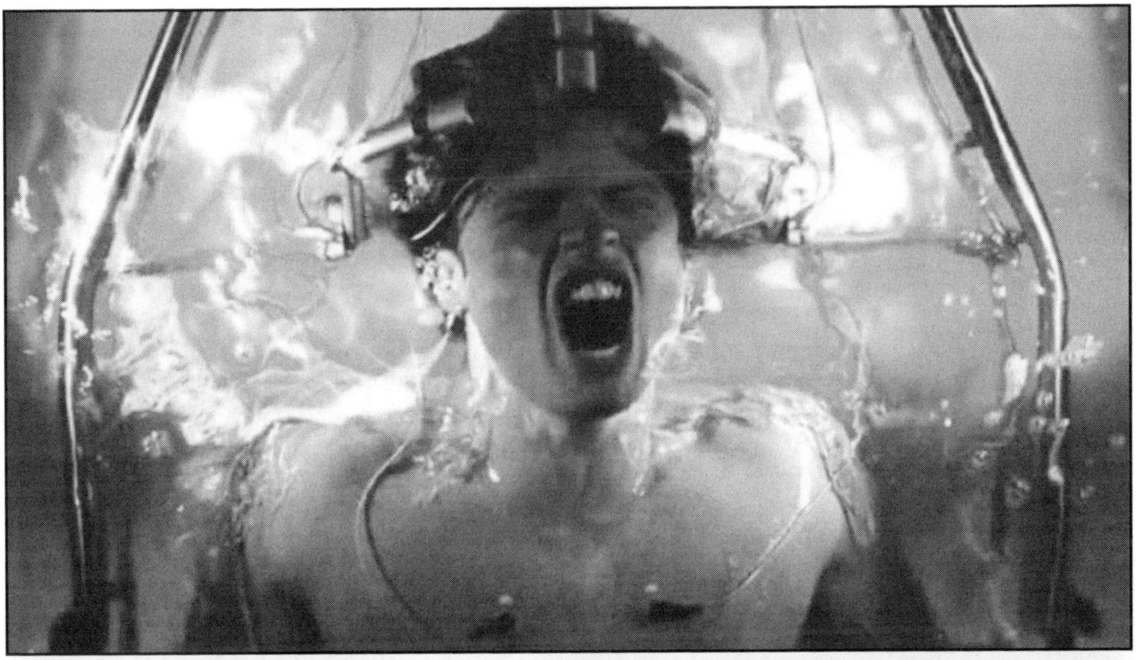

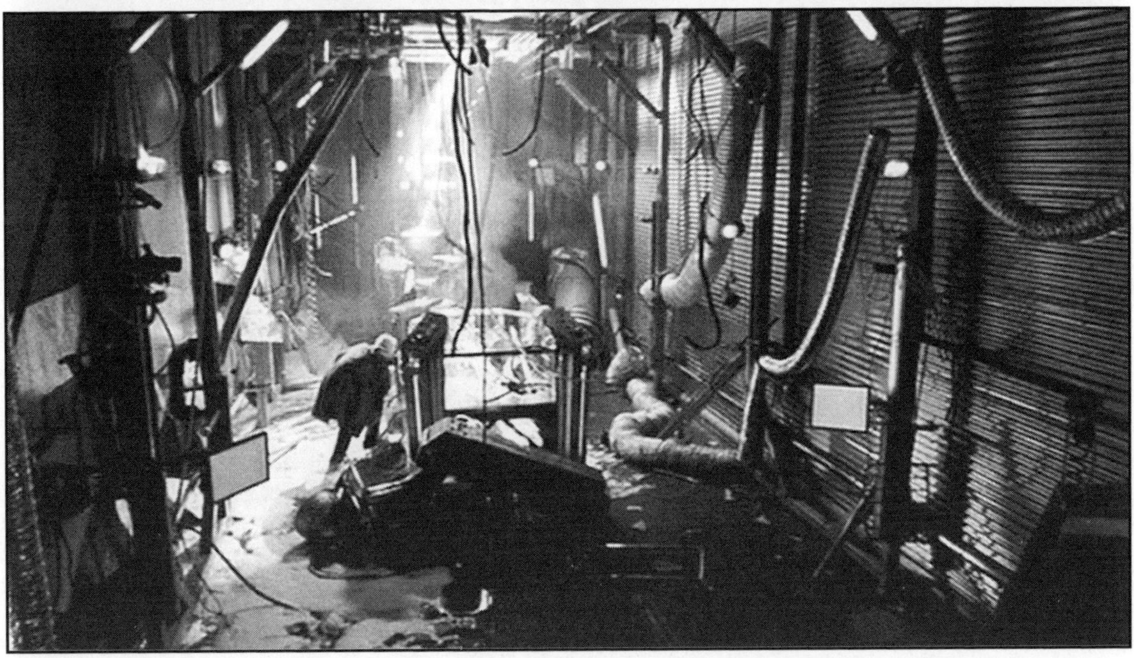

post-production any episodes that don't work, and you realize how directors screw themselves by not getting simple coverage. I had the benefit of that experience, knowing what we needed in a crunch."

One of the unusual aspects of the episode was that it didn't feature many of the regular cast except Tom Welling, Michael Rosenbaum, and John Glover. "There wasn't a need for my character in those episodes," Allison Mack says. "To be honest, that storyline stood on its own. There was no need for me in there. They were beautiful stories that didn't have anything to do with Chloe."

Although Kristin Kreuk did have one scene at the start of the episode, where she finds the modern-day Lex precariously sleepwalking, she and Allison were originally to be featured in other sequences, in which the FBI try to bug the Talon to get information out of Lex. These were cut out at the editing stage. "Anything else didn't seem appropriate," executive producer Ken Horton comments. "It was taking you out of the story. You never want to be in a story thinking you don't want to be in this scene, you want to be somewhere else."

With John Schneider absent, preparing for his own directorial début on 'Talisman', the emphasis of the story changed. "It was a show about mothers and sons," Al Gough notes. "Normally *Smallville* is about fathers and sons." Both Clark's human and Kryptonian mothers make an appearance, leading to somewhat of a problem with DC Comics. "They were developing the Superman movie," Gough explains, "And at that

Below: The footage from Belle Reve that motivates Lionel to question Clark.

Above: Lex stands on the balcony edge during his flashback.

time they were still planning an origin story. The whole discussion was about how much we could show — and in the end we couldn't show faces." The scene on Krypton borrowed imagery from a comic book retelling of Superman's origin by Jeph Loeb, in which just the hands of Clark's parents are seen.

Lex's rescue of Clark was shot at the end of a hectic day which had required the set to be distressed following the feedback from Clark's immersion and destruction of the tank. For most episodes this would be climax enough, but 'Memoria' still has two of its most important and memorable scenes to come. "The final scene between Lex and Lionel was my favorite scene to shoot of the whole third season," John Glover maintains. "It dealt with so much honesty about Lionel's relationship with him. There was a point when Lex was telling the story that I wanted Lionel to have a moment of realization of what could have been, of the mistake he made. It was a challenging scene. I love working with Michael. The set was full of people: Miles was beginning as director, so Al was there, and John Litvack. There were people there from New York which heightened the challenge."

Although the episode was originally intended to end with a shot of Lex looking at a picture of his mother, it actually finished with Clark and Martha in the barn. "That was a fantastic scene between Annette and Tom," Al Gough notes. "One of the best scenes of the series."

MEMORIA: IN DEPTH

Above: Mother and son embrace.

"That's one of my favorites," Miles Millar agrees. "Greg Beeman was sitting next to me, and he started crying. That was very rewarding, even though it was more about performance than it was about me. It was a great scene: Annette did a really good job, and it had beautiful lighting."

Millar emphasized the epic, cinematic feel of the episode, particularly for the very last image. "People don't get wide enough in TV," he says. "You want to get really wide, while most people tend to go medium or close. We don't go wide enough sometimes, so on those last shots, I wanted to go as wide as I could — it felt more cinematic to me that way."

CLARK: You're always telling me how you don't want to turn into your father, and I truly believe that. But the more you two go at each other, the more like him you become. And the more people get hurt.

The scene was accompanied by the song 'My Immortal' by Evanescence. "That song was Al's idea," music supervisor Madonna Wade-Reed says. "It was relayed to me literally as soon as the script was conceived. Al rang up and said, 'I know what song we should have at the end of that episode, I know what song I want, because

it's all about the mom.'"

Once production had wrapped on the episode, Miles Millar followed the post-production process even more closely than usual. "In my whole history with the show, that was the only time that any of the producers came from Burbank and listened to the score as I was recording it at my studio in Santa Monica," composer Mark Snow says. "In the past they had been so happy with the music and so busy with everything else that they'd never felt they had to find the time. That's one of the great things about the show for me — the interference level is zero, which is remarkable. I loved that score. There were some Jerry Goldsmith-type sounds, and that chorus effect in there. It was a little bit more avant-garde orchestral in style, rather than traditional."

Millar enjoyed the experience behind the camera. "I really want to do it again!" he admits. "Obviously the mythology episodes are the ones to do. It's a bigger piece of the puzzle, rather than just a 'Freak of the Week' or a history show. It is always gratifying to do a show in any capacity that is well-received, but to know that I wrote and directed it is always going to be a good feeling. It was such an enjoyable experience for me and people liked it — that is doubly gratifying."

"Miles was awesome," Tom Welling concludes. "I really look forward to him coming back up. The environment on set was great… it was absolutely phenomenal." ■

Above: Clark's ship prepares to leave Krypton.

CLARK KENT

"I'm just wondering if that's why I was really sent here — to change destiny. What if I make a wrong choice, and I'm not there to make it right?"

As Clark matures during the third season, his relationships with those around him change considerably. He has to come to terms with the repercussions of his adventures in Metropolis, most notably the effect on his father of the deal that Jonathan made with Jor-El in order to bring Clark back to Smallville.

"I think there's a sense of realization from Jonathan that he won't always be there," Tom Welling says, "and at least while he's here, he can give Clark a little more freedom to make some choices. They say that the important thing about raising children is to make them feel safe enough that they can leave you at some point, and feel confident enough that they can survive on their own, and I think that's what Jonathan is doing during the third season."

Tom sees similarities between Jonathan's way of dealing with Clark and Clark's way of dealing with Lana. "I guess that he's letting her do her own thing, and make her own mistakes," he says. "At the same time, he is concerned about her safety, particularly with Adam Knight. What are his motives and intentions toward her? But I don't really know what their relationship is, and maybe Clark doesn't either. If he finds himself in a situation, then he just reacts to it. If he bumps into her, it's when she's with a boyfriend. The idea of trying to think or strategize around this relationship is impossible for Clark at this moment."

The actor thinks it will be difficult for Clark and Lana realistically to get back together. "If the writers do push them back together, they're going to have to be really creative," he reckons. "I don't think that Clark can see a way that it can happen. There's so much that has gone on between them now. At the end of season three, I don't know if Clark is so much overwhelmed with the feeling of being with Lana, or afraid of losing this one person and feeling more lonely. I don't know how selfish he's being when he shows up with the flower [in 'Covenant'], then ends up disappearing after he sees Lana with Lex. The Lex variable in that situation was not something that he counted on whatsoever! He read a lot into Lex's motives, and of course Lex has a perfect answer later on about why he is there and what he is doing."

Whereas in previous seasons Clark has tended to confide in both his parents, Jonathan's heart attack, and the production logistics that allowed John Schneider time to prepare his directorial début on 'Talisman', meant that Clark had a chance for some more meaningful scenes with his mother. "Mom is mom," Tom laughs, "and they're getting a bit closer. With dad being in the hospital, I think there's been more opportunity for there to be dialogue with mom because dad isn't physically there, which then led to the whole idea of Clark not wanting to bother his dad because he's sick, so he talks to mom. I think that even though there aren't that many scenes between them,

there's more of a connection between mom and Clark this year — it's happening without the words or the screen time."

Tom describes Clark's relationship with Chloe as being "as up and down as the emotional levels between Clark and Lana! The trust issue is big between Clark and Chloe. Here's the biggest story of her life right in front of her, and she doesn't even know it. Clark is hoping that she doesn't find out, but she is constantly pursuing it behind his back, and he doesn't like it. Everyone is going behind each other's back — Lionel does it to Lex, Lex does it to Lionel, Lex is doing it to Clark, and here's Clark's best friend going behind his back when she's agreed not to."

There are, of course, extenuating circumstances. "Obviously, she's under the influence of something that she can't control," he agrees, "which we ultimately find out. 'Whisper' touches on a very big story arc between Clark and Chloe, and it's not until 'Truth', many episodes later, where it's resolved at all. And even then, it's only somewhat — there's an acceptance of each other's faults, and a promise that it won't happen again. Clark chooses to trust in that promise once more."

One of the biggest changes in Clark's life comes at the end of the season, when Pete decides to leave Smallville at the end of 'Forsaken'. Tom wasn't expecting such a dramatic turn of events. "In short, it was a surprise," Tom admits. "I always thought the idea of Pete learning the secret would then be something that would allow Clark someone else to talk to, but I think it was an interesting choice for it to be too much for him to handle, and have to leave. I think that we'll see Pete again at some point.

Obviously Pete's parents were going through some issues that he had to deal with as well, but for Clark, it's another example of how difficult it can be not only to have the abilities, but just to have the information about them."

At the end of the year, it seems as if Kara's words come true, with Clark betrayed by everyone he loves. Jonathan has made a deal with Jor-El that he has consistently misled Clark and Martha about; Lex has been maintaining his investigations into Clark despite promising that he had let the whole matter drop; and Lana has decided to go to the other side of the planet. "Clark was thinking, 'If you can't fight them, you might as well join them,'" Tom says. "And he was choosing the lesser of two evils to go with Jor-El. I think a combination of those two things would probably sum it up. A lot of times in your life, you get to a point where you go, 'I just can't fight this anymore. There's nothing I can do about it, so I better get up out of bed and go to work!' And in a sense, that's what Clark had to do. Somehow he had to try to face what it was that was causing him so much pain — and everyone else so much pain — and maybe he reasoned that by causing everyone else a little bit of pain, he could save them a lot of pain in the long run."

Tom still enjoys the process of coming to work each day, and accepting the responsibility of being the star of the show. "I realized what my role was, and how I've been put in a leadership position," he says. "Not just as Clark, but just coming in and being number one on the call sheet. You try to set a pace, but I haven't really had to do that, since the cast members come in and are ready to work, and the crew is always ready to work. If you keep your spirits up and work together with everyone, it seems to run smoothly and efficiently." ■

LANA LANG

"I don't know what I did to deserve such a great friend, but I'm glad that you came into my life."

"I think at the start of the third season, Lana had come to a place in her relationship with Clark where she was seeing them more as friends as opposed to lovers," Kristin Kreuk considers. "His leaving [at the end of season two] was a really big deal for her. Before he went, she had told him that she believed in him, and after the explosion at the Kent farm, she wanted to make sure he was okay. Then he said he had to go, without any explanation."

During the gap between the second and third seasons, Kristin decided not to concern herself unduly about the path that Lana would take during the year. "I didn't talk to the producers at all," she says. "I found that I can't invest myself too much in the show, because when I go home and I'm thinking about it all the time, it's really not healthy. I still care about Lana a lot, but I decided that I wasn't going to stress about her as much anymore. I realized it's really out of my hands, and the producers have whatever fun they have. On the occasions they've written something that just isn't the way I think she'd act, I used to get very frustrated about it, but now I just let it go. I just leave it to them, and trust them to take her in the right direction."

Kristin was as surprised as the audience by some of the revelations during the season. "I don't think that the rift between Chloe and Lana ever really healed," she says. "She knows that Chloe doesn't just want to be friends with Clark, no matter what she says. It was a big deal for her that Chloe didn't tell her she knew where Clark was, because she had let Lana suffer." However, she didn't realize that the rift was as deep as was revealed when Lana was suffering the effects of Chloe's truth drug. "I never really saw those feelings of distrust going that deep," she admits. "When she revealed that, it was strange for me, and weird to play it without the character having control over her speaking."

Lana's relationship with Lex goes through various permutations during the season. "They have an interesting relationship, and the more it develops, the more those layers come out," Kristin says. "I think Lana cares about Lex very much, mostly as a friend. I think there are undertones — more on his part than on hers at this point in time — and I think she senses that and feels that, but she doesn't think about it that much. There's something she feels that needs to have him be alright. He is a huge part of her life. He has been there for her through everything, and if she ever needed anything, she could just ask him, and it would be done in a second. He's been a very good friend and business partner to her. He's given her so many opportunities, and I think that she thinks he's a wonderful person."

That doesn't mean that Lana is blind to the less pleasant side of Lex's nature. "She

LANA LANG

can see the darkness in him," Kristin agrees, "but she can also see the light, and she focuses on that light, even though he brings out a little bit of the darkness in her."

She also isn't impressed when Lex tries to interfere with her life. "As Lana's gotten older, her relationship with him has changed a lot," Kristin says. "She respects him, but her gut instinct is to rebel against a lot of what he says, particularly when it comes to his opinion on people that she is potentially dating."

The affair with Adam Knight had originally been intended as a full-blown romance for Lana during the third year. Kristin admits that she's glad it changed. "I think the arc with Adam brought her out stronger at the end of it, which is what I really wanted to have happen at the end of their storyline together."

The violent end to her relationship with Adam is one of the factors that starts Lana thinking about moving away from Smallville, Kristin believes. "After Adam, I think she's been thinking a lot about her life and where she wants to go," she says. "By the time we reach 'Legacy', I think she's already made her decision, and I figured she's already sent in her application to Paris by this stage. She's come to a point where she realizes that she needs to take care of herself, and be okay with herself first and foremost before she can be in a relationship with anyone."

Kristin was aware that the move to Paris didn't mean that she'd be leaving the series. "They did let me know what was going on, and that my storyline was carrying on in Paris," she laughs. "Characters on our show have a knack of disappearing and taking off when you least expect them to — or dying! They're pretty good at letting you know that you're not gone if it looks like you are going."

Yet her relationship with Clark is not resolved before she leaves. "She still has feelings for Clark," Kristin says, "but she knows that she needs to really know herself better before going further with it."

Their friendship isn't helped by Lana's increasing reliance on Lex. "That's a big thing for Clark, because he's beginning to mistrust Lex, just as Lana is starting to trust him more," Kristin notes. "Clark has lots of doubts about Lex, while Lana is listening to him and respecting him, which makes it difficult for Clark. By the end of the season, she thinks she's done with him, but she isn't."

Kristin thinks that Lana might have reconsidered going to Paris if Clark had been honest with her. "If he had told her the secret, I think she would have stayed," she says. "She had a feeling he was going to tell her something of importance, something that she needed and wanted to hear. It would explain every time that he walked away from her, every time that he turned into a different person. It would explain how someone could tell you that he loves you, and then leave you as he did the previous summer. But he never told her, and I think that pushed her over the edge, and she thought, 'What am I doing? If I stay here, it'll be a foolish thing to do, and I'll hurt myself even more.'"

But even then, she asks Clark to take her to the airport. "In the end, he's her only real friend in Smallville," Kristin says. "Chloe's a friend, but there has always been the undertone of both of them liking Clark, and they've done some shady things to each other, whether under the effects of kryptonite or not. At that point, Lex has his own things going on with his father. So she cares about Clark a lot, and she'd like to see him before she goes."

Overall, Kristin enjoyed the season. "They took a different direction this year, which I think was a natural progression from the first two years." ■

LEX LUTHOR

"I will never become my father. I would never sacrifice you or anybody I care about to bring him down."

As far as Michael Rosenbaum is concerned, everything that has come before the start of the third season on *Smallville* has only hinted at what Lex is capable of. His angry outbursts, and the attack on the car that led him to meet his wife Helen were only mild compared with what was to come. "Those were really nothing," Michael says. "When Lex *really* flips out, the world needs to hide. A golf club to a car is one thing — but when he's able to really use his mind, then there's going to be trouble. Season three was the first time we really see Lex lose it. He's been facing adversity throughout the first two seasons, but has been pretty calm about it all. But then he's found true love, and it turns out to be a disaster. He's thrown onto this island and I think he really faces his worst fears. His whole world crashes down around him, and he loses who he really is."

Michael doesn't agree with John Glover's assessment that Lionel is simply testing Lex's mettle, and preparing him for the world he's going to inherit. "I think that's just bull," Michael says firmly. "There's a threshold to the abuse. Lionel is just constantly pushing Lex. Lex keeps fighting back, but he is being pushed further and further. Lex doesn't want to go there — to go to that one place that he knows ultimately he can't come back from. It's like a tug of war: once you fall down, you lose.

"But ultimately," Michael continues, "Lex is going to pull Lionel down. Every time you think they're going to reconcile, you find out something else about Lionel that you had no idea about, and you're devastated. Not only is Lex in his father's shadow, but he's also finding out how much of a monster his father is. He has to make a decision eventually, and at the end of season three, he makes the decision to put his father in prison."

It's a very tough choice for Lex to make. "The analogy would be if parents love their son, but the kid commits a murder and they're the only ones who know about it. Do they go to the authorities? I think at this point, Lex doesn't love his father anymore. He's given him so many chances and he can't deal with it any longer."

Michael isn't sure if there's one specific trigger that pushes Lex over the edge with his father. "I really think it's a lot of things," he says. "If the incident with Dr. Teng was the only thing that his father had ever done, I don't think that Lex would have sent him to prison. But by this stage, we've had 'Shattered', where Lex has been drugged, and 'Asylum', with the electroshock therapy. That was his son he was treating like that! Lionel is a monster, doing that to hide his secrets. All these things add up — killing his parents, the electroshock therapy, drugging his son, the way he treated Lex's mother — and when Lex makes his decision there's no second thought about it. In the end, it had to be done."

He admits there's a possibility that Lex is being even more devious. "There are a lot of ulterior motives in the Luthor family," he laughs. "Lex probably does have some. Perhaps by putting his father away, people will see him in a different light. At the same time, maybe they're thinking that he's a monster for putting his father away. There's an ambiguity to everything."

Michael doesn't think that Clark understands Lex. "It is hard when you're not on the inside looking out," he admits. "You haven't been put down and ridiculed for your whole life, been 'the bald-headed freak' growing up. You haven't had women come in and out of your life and screw you over, team up with your father, or try to kill you repeatedly. He's lived a horrible life, with his mother passing away because of his father and his brother being killed."

Lex's relationship with Lana alters during the third season. "There was a period in

the middle of the season where Lex was being used more for exposition," he notes, "but during that time he was finding out more about his father, and also that he innately likes Lana — platonically, or whatever you may think. I think he likes her, but I won't say anything else — I'll let the viewers decide. She's a really good person, has a good heart and means well. Lex still thinks that he's a good judge of character, and he sees something in her that's special. I can't tell you too much more, because that's all I know!"

Michael enjoys the give-and-take relationship that he has with the show's producers over Lex's progress toward the villain that he will eventually become. "The producers don't like to just give things to me," he says. "They like to give me little tastes and doses of what's going on. They give me scripts, and I read into them as much as I can."

It helps that he trusts the show's writers' instincts. "Throughout the years, I've relied on the writing, and we definitely collaborate a lot," he points out. "If something needs to be changed a little bit, we talk about it. We're building characters together and taking the journey together. Ultimately, they know where they want Lex to go, and I think I know where I want him to go. In that sense, they want me to come up with my own ideas, as well as what they're articulating on paper. They want the same direction, but with my own twists and ideas."

Along with his castmates, Michael agrees that the third season of *Smallville* is the darkest yet, "but every season Lex gets a little darker. We find out more about his past. We've come a long way with Lex from the pilot to the end of season three, and the audience must be wondering what's going to happen next. The show is about him becoming evil, and ultimately the end of the show is Clark becoming Superman, and Lex becoming the villain. When that happens, the show is over for me.

"We've had glimpses of what he is capable of, and that's what a lot of the fans are waiting for. Some love him and don't want him to turn evil; but there's the other side, with the fans who say they can't wait for him to turn evil! As an actor, it's been fun to have a little taste of that." ■

PETE ROSS

"I know that I told you keeping your secret wasn't hard. But I lied. I walk around every day afraid that I'm going to slip up."

One of the more dramatic and underplayed arcs of the third season of *Smallville* saw Pete Ross' gradual realization that he simply couldn't cope with the knowledge of Clark Kent's superpowers. As Jonathan Kent had advised him way back in season two's 'Duplicity', it wasn't an easy thing to deal with. And not only did Pete have problems with this, he also found that Clark was spending increasing amounts of time with other people rather than him. These feelings came to a head in 'Velocity' when Pete lashed out at Clark for almost abandoning him.

"We hadn't explored the difficulty of being Superman's friend," producer Mark Verheiden realized during the development of 'Velocity'. "Pete is basically saying, 'Where have you been in my life, except for saving me in 'Whisper'? When you were in Metropolis, you called Chloe, you didn't call me.' Although, of course, Dark Clark knew that Pete would have gone straight to his parents."

Another of the problems was that, as Clark would later recognize when faced with Chloe's apparent death at the end of the season, Pete was no longer really his best friend — Chloe had supplanted him. "In season one, Pete was the photographer for the *Torch*, but we never did that again," Verheiden comments. "It was a way to get him in there — but that wasn't differentiating him that much from Chloe. He was too removed from the situation. You can't do a show about who Pete is going to date today — or you could, but I'm not sure whether people would be that interested."

Sam Jones III was always pleased when the script centered on Pete. "Before I read the script, I skimmed through it," he admits. "With 'Whisper', I thought, 'Let's see how much Pete's in this…' and seeing Pete, Pete, Pete, Pete… wow, this looks like it's going to be a good one. 'Pete's come in and acting like he saved the day and he's a hero with the girls hanging off him'… That one, I went straight back to the beginning and read it carefully."

Pete's parents' matrimonial problems were the catalyst for his departure, although it surprised some people that it was enough to make him leave, and not turn to the Kents for help. "Clark really needs a guy friend," maintains Annette O'Toole. "I really loved the fact that Pete knew the secret. I wanted him there all the time; once he knew the secret Pete would be there, particularly if he had problems at home, that's the kind of family it is. I thought they would take him in, particularly now he's one of them, knowing the secret."

But it wasn't to be. "It was one of those situations where either you use him or you don't use him, and we felt sometimes that we were fighting to find scenes for him," executive producer Miles Millar admits. "We just felt that we were wasting him and wasting the character. We want to bring him back, as long as we find the right situation or storyline that will fit the character and be the prominent role." ■

CHLOE SULLIVAN

"There have been so many times when my curiosity has gotten the best of me. Maybe this is just who I am."

"I knew that Chloe was going to be working for Lionel, and I knew that she was going to be trying to get out of it," Allison Mack recalls. "That's basically been my whole storyline this year, so I guess I knew everything that happened this season. The producers kept me pretty open to what was going on with the character, and what the arc would be."

Allison wasn't surprised by any of the directions in which Chloe was taken. "I think I know her pretty well, and the writers know her pretty well, so I'm never shocked at what she does. It all makes sense, and her character growth has been pretty steady," she maintains. "The one thing I asked Al and Miles and the writers at our yearly meeting was to keep her smart. Do that, and keep her ambitious, and she'll be strong, because there aren't that many female figures out there like that for young girls. They absolutely did that, especially through the whole challenge of dealing with Lionel Luthor, and teaming up with Lex to put him away so she could save her family. I'm really happy that as she's grown, she hasn't gotten weaker or dependent, or lost any of her passion."

She considers that Chloe instantly regretted her decision to assist Lionel in the season two finale. "I think that decision was totally made out of emotion toward Clark," she says. "It was, 'You know what? You've hurt me; I'm going to hurt you.' Then the second that she took a deep breath, had a glass of water and realized what she'd done, it was a definite time for regret and wondering how she can get out of this."

Chloe realizes that she's playing in a bigger league than she has before, something that Allison had asked to see happen to her character. "The stakes are raised for her in the third season," she says. "Something we had talked about was maturing her, and a big obstacle that I thought would be interesting to watch her overcome is being in a situation where she wasn't comfortable. She's totally happy in Smallville at the *Torch* — she is in complete control of the situation there, is totally comfortable there, and never feels out of her league. She always feels she's one of the smartest people in the room — but then take her to Metropolis, or put her in a room with Lionel Luthor, in adult situations where she doesn't know how to get out of them or handle them, and she's not comfortable. She's a caged bird all of a sudden. She can't outsmart everyone."

In the end, she helps Lex and Clark to bring Lionel down. "She knows that it's make or break," Allison acknowledges. "It's the only way out for her. Her father has been fired, her life has been destroyed, and she doesn't have the *Torch* anymore, which she loves. She has nothing, so she knows she has to do it now, or never be able to break free. Lex is the only person she knows who has as much power as Lionel."

Allison thinks that the relationship between Lex and Chloe is an unusual one on the show. "There's no sexual undertone to the relationship at all, which is pretty unusual for

Smallville," she says. "Lex's only interest in her is making sure she's safe, because she's put all of her trust in him. There's an 'older brother looking out for his younger sister' vibe to their relationship. It's nice to have a male/female relationship that isn't based on sex."

As far as her ongoing feelings for Clark are concerned, Allison doesn't think that Chloe is being honest with herself when she infers to Lana that she's not concerned that Clark isn't in love with her. "She's constantly trying to convince herself that she's okay with it, but her love for Clark is so deep that she will always have the hope that he has feelings, too," she says. "She makes it easy on herself by saying, 'Oh screw it, it's never going to happen,' and maybe she's hoping it'll become a reality by saying it out loud, but deep down, it's been shown over and over that she is still hoping that there might be something. She trusts Clark, and I think a lot of that stems from the fact that she's in love with him, and she's blinded to any mistakes that he makes because of her love for him. That's something that Al and Miles and I have talked about before — her everlasting love for Clark, and the fact that she never gets over him. Her love for him and his rejection is what drives her to be as ambitious and as focused as she is."

Allison notes that Chloe begins to develop a new maturity during the third season, particularly when she takes a cold, hard look at herself as a result of the effects of the poison gas in 'Truth'. "She is completely responsible for destroying two people's lives in that story," she comments. "I think that is the first time she questions whether what she does and what she wants are right, and has to look at her own morals. It's the first time she's actually done something that's affected someone's life in the way that it did. It was a really interesting episode, and I think it caused her to step back and go, 'Wait a minute, I am Chloe Sullivan, and I am here for the truth — but with certain stipulations.' She recognizes that there's a boundary line that she won't cross — and I don't think anyone should, although that's a big problem in our society today."

The feelings for Chloe that Pete has kept hidden come to the surface during 'Truth', and there are some very touching scenes between the two characters before the latter's departure from Smallville. "I think that Pete's feelings are something that Chloe has known all along," Allison says, "but she just hasn't acknowledged. It's a very sweet and cute moment, and I'm glad that the producers put it in there before Sam went off to do other things. I hope we address it again in future — it's an important thing to happen in these two people's worlds."

Allison also enjoyed working with Sam on the interactive web/comic book adventures that formed the second batch of *The Chloe Chronicles*. "Including Sam and John Glover made it much heavier and more attached to the show," she notes.

As to future boyfriends, "We haven't had a chance as of yet to have a relationship that hasn't had death involved," Allison comments wryly. "It will be interesting to see Lex and Chloe's relationship when it's not in the utmost importance of decision-making time. Chloe has so many secrets that it would be difficult for her to have a healthy, open relationship with someone. She has that in common with Clark."

Future seasons of the show may give Allison opportunities to show other facets of Chloe, but she's delighted to have had a season-long arc. "The whole year was really about her trying to get out from under the thumb of Lionel," she points out. "It was probably the biggest arc of the show, that went all the way from episode one to episode twenty-two. There was no resolve in the middle. It was fun to take something like that and really dive into it. It was definitely a luxury to be able to go through the whole thing in-depth. " ■

LIONEL LUTHOR

"I've always tried to teach you, Lex. The devil is in the details. If you don't pay attention to the details, he'll win."

The third season saw Lionel clearly emerge as the villain of *Smallville*, as he tried to manipulate Chloe and engineered Lex's descent into madness, as well as subjecting his son to electroshock therapy. Piqued by Lex's interest in Clark, he too began to develop a fascination with the young farmboy, culminating in his attempts to get information from him by using Chloe's truth drug, and placing him into Dr. Garner's hands at the Summerholt Institute. But he didn't realize that his enemies were teaming up to deliver him to the FBI.

"It's been exciting to feel that we've made Lionel so important," Glover notes, thinking back to the almost peripheral part that he seemed destined to play at the time of the pilot. "Men of power like to be treated as important, so it's nice."

Glover particularly enjoyed his season-long sparring match with Allison Mack as Chloe. "We have such a good time with each other," he notes. "The scenes were always teasing at what was to come in the future. That was good, because that's what we were both hoping would happen, so I would tease Allison as much as I could in a scene. We were hinting on what we wanted to get at, but she's still in high school, so I don't know how much of what we want to do they will let us do! Lionel is a really sexual man, and sex comes with power. There are two stories that go through *Smallville*," he continues, "one is the complete action of that episode, with the guest stars, and then the regulars have the arc parts. That's why I think the show is so wonderful, because of the arc storylines."

His other regular sparring partner is Michael Rosenbaum. "Playing with Michael is always great," Glover says. "He's so damned inventive. I've been so lucky that he's my acting partner most of the time — it's stunning to go and play with him."

Glover thinks that Lionel simply took advantage of Lex's weakened state when he returned from the island, rather than having a major game plan to turn his son insane. "In hindsight, I'd like to say he did, because I like to think that Lionel is the most insightful and intelligent man in the world," he laughs, "but when we were shooting those scenes, I had no idea what was going to happen later."

During Lex's time working with his father, there's an element of Lionel trying to mold Lex into a copy of him, but Glover believes that Lionel sometimes seems to have two incompatible aims — he wants Lex to follow in his footsteps, but at the same time he wants Lex to be his own man. "He wants Lex to be loyal to his father, so it's tricky," he explains. "Lionel is certainly top dog, and wants to stay top dog: he wants everything. That's what makes him so complex: he wants Lex to be the best person he can be, but also be a loyal follower. That's what is so frustrating for Lex. Lionel wants him to be a

strong, strong man who still honors his father. I guess ultimately he wants Lex to stay his son, and *not* be completely his own man, because of that subservience. When Lionel says, "Honor thy father," he really means it — Lionel does have a pretty huge ego. That conflict is what makes for good drama."

Once Lex starts to show signs of mental instability, Lionel does push and exacerbate Lex's weaknesses. "He pushed, because that's how you become strong," Glover maintains. "You can't be coddled into things. That's part of Lionel's beliefs: face things head on."

Before it was exposed in 'Memoria', Glover wasn't aware of the cause of the rift between Lex and Lionel. But he hopes that "the germination of the idea for this story came from the way that Michael and I have been playing this antagonistic relationship with such love. It was wonderful to go back to the beginning, and put all that love there. Previously, I didn't know what the problem was between Lionel and Lex, but I figured that there had to be something specific. I was so pleased to find out what it was. It seemed to fit with what I've been doing with Lionel."

One of the most striking scenes of the whole season came at the climax of 'Crisis', when Lionel reached for a gun and put the barrel in his mouth. This developed from the revelation that Lionel was suffering from an incurable liver disease. Glover was made aware of this development during filming of his scenes for 'Legacy' a month earlier, so he was able to add little touches into his performance in subsequently filmed episodes. "When I was getting into the car in 'Obsession', after Clark and his girlfriend have the tour of LuthorCorp, I made a little movement, showed a little pain in my side," he notes. "I know how carefully viewers watch the show, so I tried to stick a little teaser in."

The scene for 'Crisis' was a last-minute addition to the script, and was filmed during the main unit work for 'Memoria'. Like many people, Glover was surprised at the idea of Lionel being willing to commit suicide. "But here's what I thought," he says. "Here's this man, so willful, so strong, so intelligent, and he's just been hit with this news that

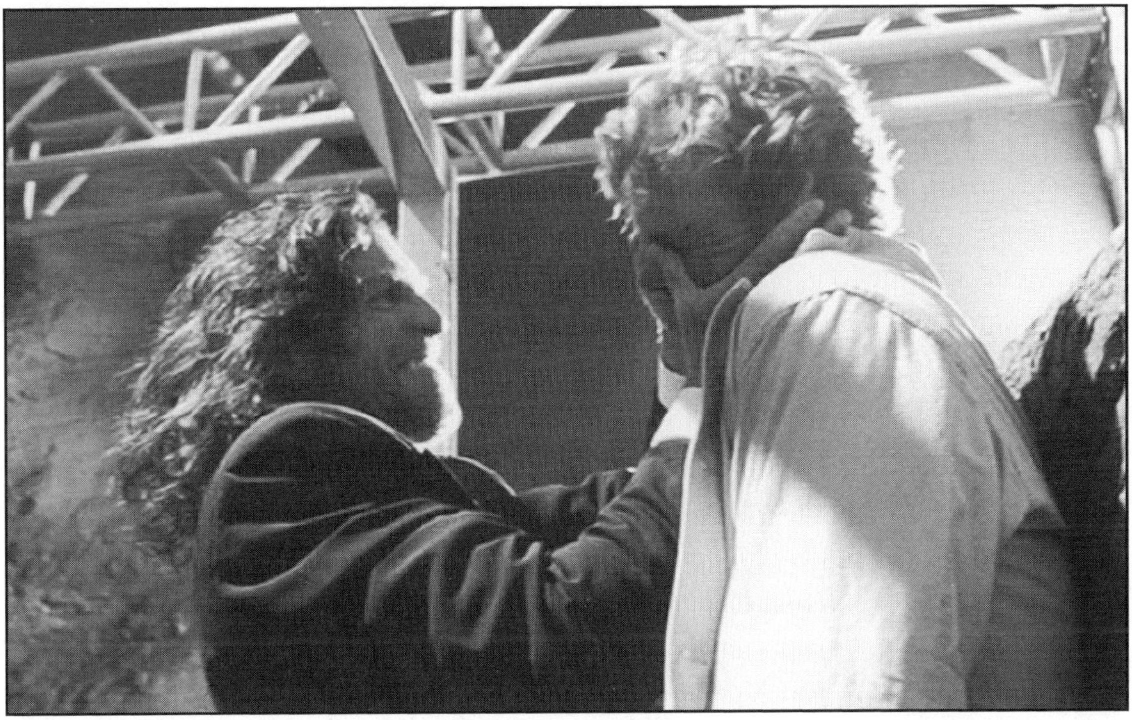

there's nothing he can do to save himself. He is human — what was he going to do? It was another contradiction that made him more interesting."

The death of one of the actor's friends also informed Glover's performance. "He was an incredible man," he relates. "Very strong, and a cheerleader to people. Instead of getting them to do what he wanted, he supported them and boosted them when they were down. Then he got cancer, and worked very positively to get over it. When he found it wasn't going to happen, he just went to bed and stopped eating. He went in about a week-and-a-half. That surprised everybody, but we'd seen a lot of friends go in very debilitating ways, and he didn't want that. How does anyone know how they'll deal with death until the time comes for you? How do you know how you're going to deal with your own death?" But Lionel's life-force becomes even stronger when he receives a phone call offering a potential new avenue of exploration. "It becomes even greater because he's gone so low that he's about to do away with himself."

Glover is intrigued by the possibilities presented now that Lionel has been arrested. "I just trust that the writers and producers will come up with some really interesting ways to deal with it, as they have the other stuff I've been involved with," he says. "They came up with such great stuff in the second season, and they certainly did again in the third, so I trust they will." ■

MARTHA KENT

"How could you make a decision like that about our son without talking to me?"

Compared with the two previous seasons, the third year of *Smallville* was a comparatively quiet time for Martha Kent, as she dealt with the aftermath of her miscarriage, and then tried to cope with the Kent farm after Jonathan's heart attack. At the end of season two, after the scenes in which Clark blew up the spaceship, causing the truck crash that led to Martha's miscarriage, Annette O'Toole had some specific requests for the producers about Martha's reactions during season three. "By doing what he did, blowing up the spaceship, Clark was trying not to cause his parents worry," she points out. "This was about him becoming a man. I said to Miles and Al that Martha can never blame Clark for what he's done. She's a mother. He has been this gift to her, and he is the only child that she will love."

Annette thinks that Martha moved on from this as quickly as she could. "I think she put the miscarriage behind her," she says. "Clark was gone for three months, and I haven't thought a lot about what she did during that time, but basically she was trying to find her son. She wouldn't dwell on the miscarriage because the Kents are trying to run a farm without the amazing help that Clark has given them. They're trying to do the everyday stuff, and cope with the finances. The one thing that she never worried about was whether he was okay or not. She knew that he was under red kryptonite, and nobody knows the effect of that on his body, but then again, nobody could harm him. It's not like what other parents go through when their kids go missing — she didn't have to wonder if he was alive or not. Then they heard about all the robberies going on in Metropolis…"

She suspects that rather more was said to Clark after he returned from Metropolis about his behavior, and the consequences of it, than was shown. "He's out there running rampant, buying cars with cash, and I don't think we dealt with all of that onscreen," she says. "There must have been more than one line. Knowing the Kents to be such straight arrows, they would make him return the money, and even if he couldn't own up, so that people didn't learn the secret, maybe they sent him to do community service or something — although, of course, he is saving people's lives every day."

Annette was delighted that her real-life husband, Michael McKean, was cast as Perry White early in the season, and she admits that she wondered "if they would write me into a scene where I could walk in and Perry just gives me a look that said, 'Oh, Mrs. Kent's kind of hot.' It would have been an in-house joke that everybody would have gotten and enjoyed. But we were just at opposite ends of a driveway out at the farm one day — that was as close as we got on camera."

For Annette, season three saw a year of being reactive, rather than proactive, so she wasn't too concerned about knowing in advance what was going to happen to her.

"If you'd asked me in season one whether I wanted to know what was going to happen, I would probably have said yes," she admits, "but in life, you don't know what is going to happen, so it's not so important that we don't know everything." ■

JONATHAN KENT

"I've lived a full life, son! Jor-El can do whatever he wants to do to me, as long as you're free!"

"I think that the relationships between the characters in *Smallville* have changed a lot, as well they should," John Schneider says of the third year, and in particular the relationship between Clark and his parents. "We're watching these characters grow together, and we're watching them grow apart."

It was a dramatic year for Jonathan Kent, beginning with his concern over Clark's disappearance, and ending with the revelation of his deal with Clark's birth father, Jor-El, and the apparent loss of his son, this time much more permanently. In the middle of the year, he suffered a heart attack, the repercussions of which continue to reverberate.

"I had an inkling that something like that was going to come up," Schneider admits, although the producers had very definitely kept him in the dark as to the exact nature of the deal between Jonathan and Jor-El. Jonathan Kent's heart problems are one of the story strands brought to the series from the original comic books, but in this incarnation are specifically linked to the actions of Jor-El. Schneider even doubts if the triple bypass surgery that Jonathan underwent in 'Resurrection' was actually necessary. "He might not have been suffering from a heart problem," he says. "I think that could have just been Jor-El getting his attention. Oftentimes doctors go, 'This symptom plus this symptom equals this cure,' but I don't think that this time that was necessarily the right conclusion. Look at the start of 'Legacy', where Jonathan falls off the roof."

As well as its physical ramifications, the deal with Jor-El has also had an emotional effect on Jonathan Kent, as he starts to wonder if he can really trust his own judgment quite as much as he believed he could. This has led to an interesting change in the dynamic within the Kent family. "I think in many regards Clark, Martha, and Jonathan are growing apart only because Clark's growing up, so they're growing apart like parents and children like to grow apart. However, at the same time, they're growing *together* as friends, which is fascinating to be a part of. In 'Forsaken' I only had one scene, where Clark says to me that he thinks that I think he shouldn't do what he's going to do. I say that I don't, but that 'I have to tell you that I don't trust my instincts nearly as much as I used to. Let's just hope that yours are right.' That is an admission of fallibility, but at the same time, it's a tremendous testimony of faith and trust in how far your son has come. Al and Miles and the whole gang in Los Angeles do a great job with the family: it's like when you work out, you're tearing a muscle apart, but at the same time you're making it stronger. That's what they've done with all these relationships."

Schneider is pleased that his concerns about the lack of scenes early in the season showing the Kents working on the farm were addressed by the producers, and in the

JONATHAN KENT

latter half of the season, Jonathan's inability to do as much as he wants to around the farm became one of the key parts of the character. Whereas before, Schneider was worried that the sense of ongoing life on the Kent farm wasn't coming through, now "we are almost always in the act of doing something else, and we stop doing that in order to deal with Clark's problem, or whatever it is," he says. "That's no longer an issue, and it's very real there now on the farm."

The actor enjoys the show's division between scenes featuring the adult characters and those in their teens and twenties. "I'm always for more adults in the cast," he comments. "I think when the adult characters are in it, it's a show that's truly skewed for everybody, not skewed quite as young as it is when it's without us."

He also hopes for more scenes between the Kents and the younger members of the cast that highlight one of the show's strengths — its realistic portrayal of the parents as human beings, rather than objects of derision from their teenage offspring. "That's really commendable that the writing has stayed strong," he notes. "I'm hoping that as the kids get older, they'll have more opportunity to come and talk to Martha and Jonathan about different things. Certainly, we're the only ones who hold whatever key there may be into the emotional makeup of Clark. It would be nice to have those conversations."

He was also pleased that 'Legacy' featured a chance for the two human fathers on the show, Lionel Luthor and Jonathan Kent, to put forward their points of view on parenting. "I had spoken about wanting to have a conversation with Lionel about how he could possibly be raising Lex his way," he points out. "It showed up as Lionel yelling at Jonathan that my son was weak because I had raised him my way. That was really great. It wasn't quite what I had in mind originally — I thought it would be a heartfelt conversation — but it was much better than what I had in mind. That's what usually happens with the writing on *Smallville*. You say to the writers, 'Gee, wouldn't it be great if...?' and they go, 'Hmmm...' and go away, write it, and make it even better."

Schneider was delighted to get an opportunity to go behind the cameras and direct an episode, something he had been hankering to do from his earliest days on the set. During the first two seasons, he had directed little sequences for the B unit, helping the

production team to keep the show filming on schedule, but he was given a chance to do a complete episode, even if that meant that he missed out on the opportunity to be involved with Miles Millar's directing début, 'Memoria'. "I was bummed that I wasn't in that, but I was prepping for 'Talisman'," he recalls. "I'm now far more aware of the abilities of the people that I work with on *Smallville* than I have been before. The nature of the show is such that we are very involved with the special effects, and we are very involved with the sets. We don't just show up — we really are part of that farm. With 'Talisman' coming up, I had the opportunity to talk to different departments about what we could do. I had the opportunity to do a couple of years' worth of preparation with regard to how I wanted to shoot different scenes in different rooms. I've had the chance to look at all those sets from many different angles over the years, unlike a new director who comes in and looks at them for the first time."

Directing 'Talisman' gave Schneider an insight into the production process. "The scripts get better through the process, rather than worse," he says. "That's very much a rarity in television. The whole process of *Smallville* continues to amaze me. Each year we do it, I think it gets better." ■

SHERIFF ADAMS

"There is too much at stake here. You find a good vantage point. You get a clean shot, you take him out."

Sheriff Nancy Adams arrived as a new broom in Smallville after her predecessor, Sheriff Ethan, was arrested for the attempted murder of Lionel Luthor (at the end of season two's 'Suspect'). Actress Camille Matthews had auditioned for the role of Byron's mother earlier in the season, for 'Nocturne'. "I found out later that Greg Beeman remembered my audition and called me in to read for the sheriff," she says. "There were some excellent actresses who had come up from LA for the audition. They were looking awfully slick in their black leather cop outfits, and I was wearing a real dumpy parka!"

Camille had the qualities that the producers were looking for. "There was a real eccentricity about her," the actress says of Sheriff Adams' early episodes, "almost a farmer's common sense that came out in her philosophical little phrases from time to time. Having spoken to a few women policemen and getting their take on how their minds work, they're constantly evaluating situations for safety. Their point of view is very distinct from a civilian's. What they look for in the course of their day is unlawfulness and danger. They're constantly twenty steps ahead and evaluating what to do if one variable goes this way, and that one goes sideways, what they should do to deal with it."

She believes that the down-to-earth sheriff is necessary to put the more comic book-like characters into perspective. "It almost feels like there are a few characters within the show to remind the audience that we are actually in small town America," she says, "and I think Sheriff Adams falls into that. I just wish that this particular character was right a bit more often. Normally when poor old Clark comes up with some hot scoop, she totally dismisses it; and she's proven somewhat incorrect a lot of the time! I find it interesting that younger people tend to hate the sheriff, I guess because she gets in the way of the teenagers on the show, and older people tend to respect her, 'cause she's the voice of authority."

Her favorite moment in the third season came in 'Extinction'. "I like those times when she's trying to piece things together in terms of what she's been told is going on by the Kents, and what she perceives to be going on," she explains. "It's the middle of the night, she's been on a manhunt, she's heard that Clark's been killed — and they're all acting as if everything is fine. I don't know why that's my favorite particularly, but I think it's a scene that brings in everything that Sheriff Adams is."

She's also glad that it gives her a chance to be a role model. "To take a woman out of the context of anything to do with sex or glamour and just give the job to her, when it happens to be a job that is usually considered a masculine job, I think is a really bold thing for a successful TV show to do." ■

MEET THE CREW

"You will not find a more dedicated bunch of people on any set. We still enjoy one another after such a long time together. I always look forward to seeing my friends there." — Annette O'Toole

It takes literally hundreds of people to bring an episode of *Smallville* to life, from the writers, producers and support staff based at the series' Los Angeles offices, to the actors, crew, technical and support personnel at *Smallville*'s studios in Burnaby, British Columbia. Many members of the team have already been introduced in the course of the episode entries in this book, but this section focuses in more detail on the roles of some key players. It's no exaggeration to say that the show couldn't exist without all of them. Everyone has their part to play…

Jeph Loeb was consulting producer on seasons two and three before taking on the role of supervising producer for season four. The veteran writer, who had combined scripting TV, features and comics, had an effect on *Smallville* at an even earlier stage.

Below: Jeph Loeb at work.

His iconic graphic novel *Superman: For All Seasons* has the same tonality that Gough and Millar would bring to the show. "Al and Miles have been very generous in terms of my work and its influence on the show," Jeph says. "I had a meeting with them during the first season, and pitched them half a dozen ideas, but I got a call to say, 'The guys have talked, and they think that for you to write an episode this season would be a waste of your time. We want you to come on next year as a producer and writer on the show!' Al and Miles felt that one of the strengths I could bring to the show was to help in creating the stories that are broken into outlines and are then built into the scripts."

Loeb sees his job as simply "telling stories. We spent time coming up with ideas. 'This story's never been told... what characters in Smallville have we not explored — teacher, milkman, mailman, farmer?' We filled up white board after white board with all those ideas before we started pitching actual stories. It's the joke that I always say to the guys when we're meeting in the office and I have to go back to the writers' room. I say, 'I'm going to go back to making stories.' That's an area that I'm very comfortable in. I love the aspects of production, being on set and doing all that stuff, but this job is about telling stories."

The writers' room itself was run during the third season by industry veteran **Ken Biller**, who had been executive producer of *Star Trek: Voyager* during its final season. He joined *Smallville* from another Vancouver-based show, *Dark Angel*. "Miles and Al were revamping some of the writing staff from the first season," he recalls. "I had actually met with them the year before, and they told me they were looking for someone to run the writers' room for the second year. I watched or read all the first season episodes, and Al and Miles told me which ones they liked and which ones they didn't like as much, and why. There was a mandate to do more of a mix of shows. They didn't want to do a krypto-villain every week, they wanted to mix in more mythology and true crime-type episodes."

Above: Jeph Loeb and colleagues have a lighthearted 'disagreement' in the writers' room.

Although Biller was given a degree of leeway, "everything ultimately gets vetted by Al and Miles," the producer explains. "They may not be there for some period of time while the stories are being worked out, but then one or other, or both of them, will come in and I will pitch them what we have on the board, and what we're thinking about doing. They'll say, 'That's fantastic,' or 'Try something else.' It was clear to me from the beginning that they are the bosses, and for me, the more they are in the room, the better. I'd rather have them invested in what we're doing as early as possible. It's their show. My job is to help them realize what their vision of the show is. I tell new writers, 'It's not your job to come in and do a Ken Biller script, it's your job to come in and do an episode of *Smallville*. If the people who created *Smallville* tell you they want it to be something else, then you can have a passionate argument why you believe a certain thing, but at the end of the day they have a moral authority and a certain sway to say this is what they think, and that's what you do.'"

During filming, the actors rely heavily on support from everyone in the crew, but inevitably there are some with whom they come into contact more often than others. Annette O'Toole cannot speak highly enough of onset wardrobe supervisor **Steve Oben**. "He is one of the single most helpful and optimistic people I've ever known in my life. He knows everything that is going on on the set, sometimes sooner than the A.D.'s [assistant directors] do. I find his humor, energy, and insight invaluable, and in a lot of ways I think he is the heart of our little *Smallville* family. And he keeps us very warm on those cold nights on exterior locations. He gives us 'hot shots', those little packs of chemicals that warm up when you shake them, and he puts them in our pockets for us to hold. When the director calls 'cut!' he throws a big down coat over us and holds us to get us warm again. All the women on the show appreciate this very much when our teeth are chattering! He is just one of an amazing group, though. I am always in awe of the crew because of all the long hours and hard work they put in, and how little recognition they get for it. I hope they know how much the cast appreciate them, and how ridiculous we would look without them."

At the other end of production, no episode of *Smallville* would be complete without its very distinctive soundtrack. While award-winning composer Mark Snow creates the instrumental score for each episode, the other musical elements are found by music supervisors **Madonna Wade-Reed** and **Jennifer Pyken**. "Our job is to supervise the music that comes from any source," Madonna explains. "If a car radio is playing, if we're in the Talon or anywhere else where you wouldn't be without music, it's our job to find all that music, present it to the 'powers-that-be', collectively decide which pieces work best and which suck within the body of the work, and then get it cleared — do all the legal paperwork to allow us to use it."

Madonna reads the scripts as soon as they are issued, and 'red flags' anything that might be relevant. "If there's something being played on camera that will need pre-clearing, I'm involved before anyone else even sees the script," she explains. "For example, if it says that there's a party and somebody is singing 'Happy Birthday', it has

Below: The score emphasises a climactic moment in 'Covenant'.

to be cleared. It's not in the public domain."

If there aren't pre-clearances, the music supervisors flag the pages of scenes that might benefit from music. Those are then discussed with the music editor, Chris McGeary, and with the producers. "It's Al and Miles that you've got to plead your case to if you really want a song," Madonna notes. "At the end of the day they make the final decision."

Often the executive producers will indicate a specific song that they want used, as in 'Memoria'. The music supervisors will present their options, which are used in the editing process while clearances are being sorted. "*Smallville* has such a strong appeal across the board that record labels recognize that it is a means of getting the music out there," Madonna says. "Sometimes we get denied, and then it's our job to get a replacement. Occasionally bands will feel that a song has been overexposed. We have to go back to the producers or director and say that this time we can't honor their trust."

Music editor **Chris McGeary** is the link between the producers, the music supervisors, and the composer. "Once a show is near completion, the producer, the director, the composer and I will sit down and figure out what kind and what style of music we want, and where we want it to be," he explains. "I take copious notes for Mark, and we figure out where and how the music should be played. The producers talk in broader strokes about what they want — they want a climax to hit when the door closes, or set up a mood and let it peak during the next couple of scenes. We don't talk very much in terms of instruments, because they know pretty much what the pallet of the show's sound is by now."

More memorable musical moments at the stunning ends to two major episodes — 'Memoria' (below) and 'Covenant' (opposite).

From those notes, Mark Snow has about five days to compose anything up to thirty-five minutes of music. "Mark sends it to me, I mix it, and then cut it using a Pro Tools program which is then brought to the mixing stage," Chris continues. "After we lay the music in and it's all mixed, we have a playback for the producer and director, and if they don't like what Mark has done and want it changed, then I do some editorial work right there on the stage to make it right. I do the same with the songs that Madonna and Jennifer find: they clear them, but I'm the one who has to cut them to make them fit the scene."

The relationship between the score and the songs differs from episode to episode. "Sometimes we have songs which are almost played as score, and take over the whole scene," Chris says. "Other times, we want the song to stand out, so Mark will write the score so that he clears some room so that the song can really pop. A lot of times, we have to segue from a score piece into a purchased song, so we try to get them to Mark as early as we can. We have to make sure the key relationship is correct, to make sure that the song doesn't jar."

As with every aspect of *Smallville*, the final decision rests with the executive producers, but they give their people the license to do their jobs. "It's my call on the stage," McGeary says. "I tell the mixers what level I want the songs or the score to be played. We're all proud of the job we do on *Smallville*, and we all try to do our best. It starts at the very top with Al and Miles — they've been great to everybody, and they put 110 percent into it. They expect the same, and we're glad to give it to them." ■

THE CHLOE CHRONICLES

"You're fishing in deep waters, Miss Sullivan. Be careful you don't hook a monster."

Following the success of the *Chloe Chronicles* promotional Webisodes during the course of the second season, Mark Warshaw and his team worked not only on a second batch for the third year, but also created a series of commercials (in conjunction with the cell phone company Verizon) to bridge the gap between seasons. For American audiences, by the time the season première 'Exile' aired, they were already aware of some of Clark's exploits in Metropolis.

There were five commercials in all. In 'Bullet', a skater called Kyle tells reporter Max Taylor about a 'big dude' who attacked an ATM and stole $5,000 or more. Max tells his colleague Tammy that he is sticking with his nickname of 'Urban Legend' for the thief. In 'Chance', Tammy tells Max that a new columnist, high school girl Chloe Sullivan, thinks that the Urban Legend might be linked to a small town called Smallville. In 'Practice', Max hears more testimony, this time about the Urban Legend's ability to leap buildings in a single bound, going from the top of the *Daily Planet* offices to the roof of the LuthorCorp building. In 'Locomotive', the Urban Legend's activities are becoming more blatant, as he throws the globe from the top of the *Daily Planet* building into a storefront, and an eyewitness notices that his eyes are glowing red. In the final installment, 'Tied', Metropolis Police Department detective Maggie Sawyer (seen in 'Insurgence', and a regular character in the comic book continuity of Superman) shows Max some more of the Urban Legend's

Right: Chloe Sullivan on the trail once more.

handiwork — a parking meter that has been used to tie up and restrain someone whose bike the Legend has stolen.

"That was something that the network got involved with, in a way to get more promotion for the show," executive producer Al Gough notes. "We were involved with Mark Warshaw in the conception of the ads. We wanted to make sure that the ads had the tone of the show. We were very pleased with them to the point where we used the guy in the ads, Max, in 'Delete' — and killed him soundly with a pencil! That was an interesting promotion, playing on what Clark did in his summer in Metropolis. The show is very expensive to produce and these are opportunities to put more money into the show and do bigger events."

Above: Coming out from behind the camera — Pete and Chloe discuss the clues.

A few months later, Warshaw reunited with Allison Mack to create a second series of *The Chloe Chronicles*. This time, much to Allison's pleasure, the scripts also involved Sam Jones III and John Glover, with Pete Ross revealed to be her partner filming the episodes, and Lionel turning up at the end. These innovative new *Chronicles* didn't only involve the Web audience — the plotline also tied in to the *Smallville* comic book from DC Comics. While issue 7 bridged the gap between the first and second promotional series, issue 8 picked up from the cliffhanger to the sixth part of the *Chronicles*, recreating the final frames of the episode as the splash page of the book, and finishing itself on a cliffhanger, which was then filmed in live action as the start of the final episode on the web.

The comic book segment in particular tied in heavily to the show's continuity, with various of the krypto-freaks who had apparently died, including Greg Arkin, the Bugboy from season one's 'Metamorphosis'; Sean Kelvin, Chloe's boyfriend from 'Cool'; and the late Tina Greer, last seen menacing Clark in 'Visage', seen in the underground lab that Dr. Walsh has been working in. A message from Walsh sets Chloe and Pete on the trail, and their investigations lead them to the lab where they encounter Dr. Jamison, who is carrying out further experimentation on the krypto-freaks. Jamison is suffering from the jitters caused by prolonged exposure to kryptonite, and knocks into a propane tank, which explodes. As Pete and Chloe escape from the scene, they are met by Lionel, who coincidentally is nearby.

"It's a really cool, really new sort of idea of going from page to screen," Allison explains. "Using Sam and John Glover made the second run much weightier, and more attached to the show. We're getting more of a full thread of storyline building all the way through. It's developing into this actual show that could stand on its own. It's been fun — I've had a really good time doing it, and I'm really lucky that Mark had the idea and wanted to work with me." ■

THE PHENOMENON

"I love *Smallville* and won't miss an episode, and it's every bit as influential as the rest of Superman's grand history." — *Superman: Birthright* writer Mark Waid

"*Smallville* works because it plays like a comic book come to life: it's grand, mythical, and action-packed. The show is also unique in that its viewers know what happens beyond the series. We know Clark has to return to the good side, so he can grow up geeky and awkward and eventually work in the newsroom of *The Daily Planet*. We know Lex Luthor becomes Superman's nemesis. What we get here is a glimpse of a familiar journey with a preordained conclusion, so the show is all about the struggle to get there."

So wrote Mary Colgan in a review of the start of the third season of *Smallville* that appeared in the online magazine *PopMatters* in November 2003. While some mainstream journals seemed to find it difficult to acknowledge that the show had moved on from its 'Freak of the Week' roots, it was clear to the fans and the general audience that season three marked a step up from the previous years. Interest in the show was increasing, with many of the genre entertainment magazines giving prominence to coverage of the series, even as preparation was well underway for the licensed magazine and books from Titan.

Above: Michael McKean (Perry White) with real-life wife Annette O'Toole (Martha Kent).

To coincide with the launch of the new season, Tom Welling and Kristin Kreuk went on the round of chat shows, promoting the new episodes, in particular the changes in Clark's character seen during his time in Metropolis. Combined with the Verizon promotional campaign, this led to a heightened awareness of the series, which was added to by the publicity that surrounded the appearance on the show by *This Is Spinal Tap*'s Michael McKean as Perry White a few weeks later.

Smallville continued to encroach on the public consciousness during the fall of 2003. Popular comedy series *Will & Grace* made a highly complimentary reference to the show in an early episode of its new season, in particular praising the quality of the writing. The prestigious glossy journal *Architectural Digest* carried a feature on the show's standing sets. *Entertainment Weekly* contained a major article which suggested various ways that the show could continue to improve: including bringing Clark and Lex into conflict, making Chloe evil, and bringing in more references to Clark's future as Superman.

With a new Superman movie continuing to be discussed, there was considerable fan speculation as to whether Tom Welling might be persuaded to don the Superman costume for the new film. Tom himself confirmed that he had had a meeting with the then-director, Brett Ratner, but that nothing had come of it. Highly respected artist Alex Ross followed his 2001 *TV Guide* painting of the *Smallville* cast with a speculative portrait of a slightly older Tom in the Superman costume, as if to coincide with the

release of the movie in 2007. There was also speculation as to whether Bruce Wayne might appear on *Smallville*, with some conjecture that Adam Knight might really turn out to be an alias for the Caped Crusader, but this rumor was laid to rest with the announcement of filming on *Batman Begins* in mid-2004.

The fans watched the show closely, picking up the little details that were hidden in the background. Newspaper headlines in 'Asylum' referred obliquely to Wonder Woman, while the license plate number of the Dukes of Hazzard's car, the General Lee, turned up as part of Martha Kent's email address in 'Delete'. The series' fans also showed their appreciation in more practical ways, supporting John Glover's Memory Walk for sufferers of Alzheimer's disease.

Smallville's influence, both within the TV industry and on the comic books from which it sprang, was recognized in the spring of 2004. The annual William S. Paley Television Festival, sponsored by the Museum of TV and Radio in Los Angeles, hosted a seminar which allowed the public to pose questions to the producers, as well as many of the cast and crew. Meanwhile in the DC Comics universe, 2004 saw a further retelling of Superman's origin in Mark Waid and Lenil Yu's *Birthright*, which, for the first time since the Silver Age, featured Lex and Clark growing up together in Smallville, as they were now doing in the TV series. Writer Waid said that he would have wanted to put young Luthor back in Smallville anyway, but "I will cop to the fact that it was *Smallville* that showed me with stunning clarity how cool it would be to have Ma and Pa Kent actually be parents, not kindly, white-haired grandparent types."

As the season came to a close, the rumor mill started to work overtime that there would be new characters joining the show in the fourth season — not least, someone who would become probably the most important person in Clark Kent's life. With the show's creators revealing that the new year would see a return to the show's high school roots, the stage was set for a very different fourth year... ■

Opposite: The cast for season four, including Erica Durance as Lois Lane, and Jensen Ackles as Jason Teague.

22 Things I Learned This Year

By Chloe Sullivan

Lesson 1: Be careful helping those in exile — friend or foe, self-imposed or not.

Lesson 2: The phoenix always emerges from the ashes more powerful than ever.

Lesson 3: Even meteor freaks deserve a fate better than extinction.

Lesson 4: While in slumber, pay close attention to imagery of socks. (It's somewhere online... If you don't know what I'm talking about, better ask somebody.)

Lesson 5: Something tells me this is not the last we've seen of the man called Perry.

Lesson 6: A relic can give rise to old memories.

Lesson 7: Magnetic attraction can result in two weeks' probation.

Lesson 8: Bad parenting can result in a shattered mind.

Lesson 9: Fact: There are more former SHS students at Belle Reve Insane Asylum than at any one university in the US. It ain't the water here, kids.

Lesson 10: If you are inclined to whisper it, you probably shouldn't say it.

Lesson 11: If you don't know who sent it... delete it.

Lesson 12: Remember the butterfly; don't mess with the hereafter.

Lesson 13: Minus Pete Ross, the days will pass with a much slower velocity around here.

Lesson 14: We all have an obsession (I know I do), but you have to know your boundaries — even if you think you are hotter than Lana Lang.

Lesson 15: Resurrection should be left to a higher power, not a biologist.

Lesson 16: The Chinese character for the word crisis is the combination of the symbols representing extreme danger and extreme opportunity. Think about it.

Lesson 17: Leave a legacy. Make it positive. A purposeful life is a life of purpose.

Lesson 18: As much as I love the truth — it can hurt.

Lesson 19: Sometimes the only way to describe a beautiful thing is to coin a word like memoria.

Lesson 20: A talisman is the property of the culture that created it, not the "looters" who looted it.

Lesson 21: When forsaken, a child will grow desperate for affection.

Lesson 22: Do not break a covenant.